SpringerBriefs in Applied Sciences and Technology

Reliability

Series Editor

Cher Ming Tan

For further volumes:
http://www.springer.com/series/11543

Cher Ming Tan · Feifei He

Electromigration Modeling at Circuit Layout Level

 Springer

Cher Ming Tan
Feifei He
School of Electrical and Electronic Engineering
Nanyang Technological University
Singapore
Singapore

ISSN 2196-1123   ISSN 2196-1131 (electronic)
ISBN 978-981-4451-20-8   ISBN 978-981-4451-21-5 (eBook)
DOI 10.1007/978-981-4451-21-5
Springer Singapore Heidelberg New York Dordrecht London

Library of Congress Control Number: 2013931512

Printed on acid-free paper

Springer is part of Springer Science+Business Media (www.springer.com)

Preface

Electromigration (EM) has been a dominant failure mechanism for integrated circuits' interconnections since the beginning of integrated circuits (IC). Extensive studies on electromigration, both theoretical and experimental, have been done in the past and modeling of electromigration that enable us to have a better understanding and prediction of the electromigration weak spots and time to failure are also developed. However, almost all the previous works are done on the test structures instead of at the circuit level, and as such, they are useful for the evaluation of interconnect technology in the wafer fabrication. With the increasing interaction between circuit performance and materials that make up an integrated circuit as we advance in IC technology nodes, will there be a difference in the electromigration behaviors between the test structure and circuit level? This book serves as an opening to address this question, and hopes to inspire more research to be done in this area.

In particular, this book aims to model the EM reliability of interconnects at circuit layout level using 3D model. In order to perform the modeling, a method to construct a 3D finite element (FE) circuit model from a given 2D IC layout must be developed so that the transient electro-thermo-structural simulations using both Cadence (a circuit simulator) and ANSYS (a finite element software) can be performed. In this book, a simple two-transistor circuit with two metal layers is used as the example to illustrate the method of such construction.

Once a 3D FE circuit model can be constructed for a given IC layout, and with the FE simulation, the EM weak spots of the interconnects in the IC can be analyzed to study the effects of current density, temperature gradient, and thermo-mechanical stress gradient on the EM reliability of an IC. With this analysis, one can compare the EM weak spots of a 3D circuit structure and a standard line—via test structure under both the EM test condition and the circuit operation condition to answer the question that we posed at the beginning, and the answer is Yes.

With a 3D FE circuit model, we can now examine how we can improve the EM reliability of a circuit based on the circuit model instead of the test structure. Along this rationale, we investigate the effects of barrier layer thickness, dielectric material, layout, and process modifications (e.g., interconnect structures, transistor placement, stress-free temperature of the metallization) on the EM reliability of integrated circuits, without altering their functionalities and violation of circuit design rules.

It is our hope that this work can push the EM research from test structure to the "system level" so that the research outcome can have a more direct benefit to the circuit designers, leveraging on the maturing of the 3D EM models on test structures that have been reported in the literatures and the increasing computational power of modeling software. This is especially needed as we advance in the IC technology node where interaction between materials, devices, and circuits can no longer be neglected.

Contents

1 Introduction   1
  1.1 Overview of Electromigration   1
  1.2 Modeling of Electromigration   2
  1.3 Organization of the Book   4
  1.4 Summary   4
  References   5

2 3D Circuit Model Construction and Simulation   7
  2.1 Introduction   7
  2.2 Layout Extraction and 3D Model Construction   7
  2.3 Transient Electro-Thermo-Structural Simulations
     and Atomic Flux Divergence Computation   12
     2.3.1 Transient Thermal-Electric Analysis   12
     2.3.2 Transient Structural-Thermal Analysis   20
     2.3.3 Application of Submodeling   22
     2.3.4 Computation of Atomic Flux Divergences   23
  2.4 Simulation Results and Discussions   28
     2.4.1 Current Density and Temperature Distributions   28
     2.4.2 Transient Temperature Variation During
         Circuit Operation   31
     2.4.3 Effect of the Substrate Dimension on the Circuit
         Temperature   33
     2.4.4 Transient Thermo-mechanical Stress Variation During
         Circuit Operation   34
     2.4.5 Distributions of the Atomic Flux Divergences   35
  2.5 Effects of Barrier Thickness and Low-κ Dielectric
     on Circuit EM Reliability   40
  2.6 Summary   44
  References   45

3 Comparison of EM Performances in Circuit
   and Test Structures   49
   3.1 Introduction   49
   3.2 Model Construction and Simulation Setup   49
   3.3 Distributions of Atomic Flux Divergences Under Different
        Operation Conditions................................   52
   3.4 Effects of Interconnect Structures on Circuit EM Reliability....   63
   3.5 Effects of Transistor Finger Number on Circuit
        EM Reliability   68
   3.6 Summary   73
   References   74

4 Interconnect EM Reliability Modeling at Circuit Layout Level ...   75
   4.1 Introduction   75
   4.2 Model Construction and Simulation Setup   75
   4.3 Distributions of Atomic Flux Divergences   79
        4.3.1 Total AFD Distribution of the Full Model   79
        4.3.2 Total AFD Distribution of the Sub-model   84
   4.4 Effects of Layout and Process Parameterson Circuit
        EM Reliability   88
        4.4.1 Line Width and Degree of Turning...............   89
        4.4.2 Transistor Orientation   91
        4.4.3 Inter-Transistor Distance   93
        4.4.4 Stress-Free Temperature of the Metallization   97
   4.5 Summary   97
   References   98

5 Concluding Remarks   101
   5.1 Conclusions   101
   5.2 Recommendations for Future Work.....................   103
   References   103

Symbols

h	Convection heat transfer coefficient
V_{dd}	Supply voltage
V_{in}	Input voltage
V_{out}	Output voltage
V_{bias}	The bias voltage
I_{ps}	Current flow from the V_{dd} line to the source of PMOS
I_{pd}	Current flow from the drain of PMOS to the output line
I_{nd}	Current flow from the output line to the drain of NMOS
I_{ns}	Current flow from the source of NMOS to the ground
I_{out}	Output current
δ	Skin depth
ω	Angular frequency of current ($2\pi \times$ frequency)
μ	Absolute magnetic permeability of the conductor
V_t	Threshold voltage of the transistor
t_f	Time-to-failure
V_c	Critical volume of mass depletion or accumulation required for failure to occur
Ω'	Atomic volume
p_w	Effective diffusion path width
p_t	Effective diffusion path thickness
$\nabla \cdot J$	Total atomic flux divergence
J_{EWF}	Atomic flux due to electron wind force
J_T	Atomic flux due to temperature gradient induced driving force
J_S	Atomic flux due to thermo-mechanical stress gradient induced driving force
N	Atomic concentration
k_B	Boltzman's constant
T	Temperature of the metal line
eZ^*	Effective charge of the diffusing species
j	Current density
ρ	Resistivity of the conductor, which depends on temperature as $\rho = \rho_0(1 + \alpha(T - T_0))$

α	Temperature coefficient of resistivity
ρ_0	Electrical resistivity at temperature T_0
D_0	Prefactor of the self-diffusion coefficient
E_a	Activation energy
Q^*	Heat of transport of the metal line
σ_H	Local hydrostatic stress
∇T	Temperature gradient along the metal line
$\nabla \sigma_H$	Stress gradient along the metal line
E	Young's Modulus of the metal line
α'	Thermal expansion coefficient, CTE, of the metal line
v	Poisson ratio of the metal line
$\nabla \cdot J_{\text{EWF}}$	Atomic flux divergence due to electron wind force, AFD_EWF
$\nabla \cdot J_{\text{T}}$	Atomic flux divergence due to temperature gradient-induced driving force, AFD_T
$\nabla \cdot J_{\text{S}}$	Atomic flux divergence due to thermo-mechanical stress gradient induced driving force, AFD_S
W_t	Width of the transistor

Chapter 1
Introduction

1.1 Overview of Electromigration

With the continuous increasing circuit complexity and the down-scaling of the minimum feature size in integrated circuits (ICs), the electromigration (EM) reliability of ICs is becoming increasingly important. As re-design and re-manufacture are very time- and resource-consuming for present-day, ultra large-scale integration (ULSI), "Design-for-Reliability," is essential.

Reliability of an IC depends on the reliability of the embedded circuit elements, including the transistors and the interconnects. The interconnect dimension scales about 30 % with each advancing technology node, and this causes an approximate 1.5 times increase in the maximum current density and an increase in the ratio of the Cu/cap interface (i.e., the fast Cu diffusion path) to the total Cu volume [1]. In present-day ICs of nanometer scale, the interconnect reliability has become the main factor that determines the circuit reliability [2], and the failures in the interconnects are mainly due to EM. The study by Srinivasan et al. [2] showed that the failure rate of a processor using 65-nm process is more than 3 times higher than that using 0.18-μm process, and the main cause of failure was EM. This is because a narrower interconnects tend to have higher current density and are more sensitive to the increase in the interconnect line resistances due to voids formation, especially when the circuits are operating at higher frequencies [2]. Therefore, they are expected to have a shorter time to failure and a higher failure rate. Furthermore, the poor thermal conductance of the low-κ materials aggravates the EM deterioration [3]. Therefore, EM turns out to be the most dominant failure mechanism in a high-density ICs with highly scaled interconnect dimension [4], and it is important to ensure a long EM lifetime if the reliability of an IC is to be maintained or improved as we advance in the IC technology node.

EM is the gradual displacement of metal atoms in a metal line when the current density is sufficiently high (above 10^6A/cm^2) that causes the metal atoms to drift in the direction of electron flow [5]. The number of atoms passing through a specific cross-sectional area in a unit of time is called the atomic flux [6]. The difference between the atomic flux into and out of a volume element per unit time is the

C. M. Tan and F. He, *Electromigration Modeling at Circuit Layout Level*,
SpringerBriefs in Reliability, DOI: 10.1007/978-981-4451-21-5_1, © The Author(s) 2013

atomic flux divergence (AFD), and it is the main cause for EM failures in IC interconnects [7]. Voids or hillocks are formed when the number of atoms arrives in a volume element is different from the number of atoms that depart (i.e., when there is a non-zero flux divergence). The formation of the voids and hillocks can cause open circuit or short circuit and hence results in serious reliability problem.

The rate of the atomic flow depends on the magnitudes of the force that tend to hold the atoms in place and the force that tend to displace them [5].

The factors that affect the "holding" force include the nature of the metal, its grain size, orientation, and grain boundary, as well as its interface with other surrounding materials. These factors are mainly the physical properties of the metal line [5]. The "holding" force is larger in a metal line with higher self-diffusion activation energy, lower resistivity, smaller number of diffusion interfaces, and lesser amount of grain-size variation [7]. A large "holding" force is favorable in its EM lifetime as the metal atoms are more resistant to the atomic flow.

The factors that affect the "dislodging" force include current density, temperature, and thermo-mechanical stress [5]. High current density can greatly accelerate the EM degradation by displacing more metal atoms at a faster rate [7]. When the current flows through a metal line, heat is generated due to Joule heating [8]. The rise in temperature not only increases the rate of the atomic flow but also causes a non-uniform temperature distribution, especially in long metal lines with complex structure and/or material inhomogeneity [4]. The non-uniformity in temperature can result in temperature gradients in the metal line. These temperature gradients render the metal atoms to flow from the high-temperature region to the low-temperature region [7]. The temperature gradients also induce thermo-mechanical stress due to the thermal mismatch between the metallization and its surrounding materials. The magnitude of the thermo-mechanical stress is affected by the surrounding materials and the stress-free temperature of the metallization (i.e., the temperature at which the stress in the metal line is zero, usually taken as the final annealing temperature of the metallization) [7]. The metal atoms move along the metal line in order to relax the thermo-mechanical stress, and this movement is faster when the stress gradient is higher.

As all of the above-mentioned factors affect the EM process in the interconnects, it is important to include all of them when performing EM modeling of the interconnects. For comprehensive understanding of the EM physics, one may refer to [7].

1.2 Modeling of Electromigration

Simple line or line-via test structure is generally used in EM test to assess the reliability of interconnect technology in semiconductor industry. Thus, EM modeling is usually performed on such test structures. 3D physics-based EM models [9–12] have the advantages of including all the above-mentioned factors. These models can predict the EM lifetime and failure sites of the interconnects [9–11], simulate the void nucleation and growth process [12], and evaluate the back stress, microstructure, and material effects [12]. However, the model based on the simple line or line-via test structure may not always be able to predict the correct EM failure sites and is not a suitable choice if we want to study the EM reliability of an entire IC under the circuit operation condition, as will be shown in this book. To investigate the EM reliability of an IC, the entire interconnect structures must be considered as a part of a system that includes the package and the Si substrate. This drives the need for 3D EM modeling at circuit layout level.

Many circuit layout level EM simulators had been reported in literature, and all of them are 2D. The traditional 2D circuit simulators [13–16] assumed a constant temperature across the surface of the chip. This assumption is no longer valid at the 90 nm technology node. In high-density circuits with multiple metal layers, the temperature difference at different portions of the chip may vary by 50 °C or more and that between the topmost and the lowest layer of the metallization can be larger than 45 °C [17]. The temperature effects can significantly affect both the performance (e.g., timing, power consumption, noise) and the reliability (e.g., lifetime, failure rate) of an IC [17].

As explained earlier, the rate of the atomic flow is faster at higher temperatures. Furthermore, the material degradation rates of both the metallization and its surrounding layers are faster when the temperature is higher [18]. Therefore, the temperature effects cannot be ignored, and the later EM simulators [19–22] took these effects into consideration. However, due to the limitation in dimensionality (only 2D), these simulators were unable to correctly represent the true temperature profile of the circuit as the heat dissipation of an IC depended on its actual 3D structures [17]. It is important to include the full 3D temperature profile of the circuit into the simulation tools so as to correctly assess the thermal impacts on the chip [17]. Hence, a 3D circuit model for an IC is needed as its actual physical implementation in a wafer is indeed 3D in nature.

Also, all the 2D circuit simulators ignored the thermo-mechanical stress effects. The work by Li et al. [23] showed that current density was no longer the sole driving force for EM when the interconnect line width became less than 0.20 μm. In fact, the thermo-mechanical stress was the dominant factor in determining the EM reliability of the interconnects and should not be ignored. Moreover, after the formation of the voids and hillocks, the presence of the back stress in the interconnects can relief the EM degradation and thus affect its EM lifetime.

The temperature and thermo-mechanical stress in various parts of an interconnection in IC are highly dependent on the surrounding materials and their material properties (e.g., thermal conductivity, thermal expansion coefficient, Young's modulus, Poisson ratio). The ITRS for interconnect [24] pointed out that with the use of various interconnect, barrier, and dielectric materials, it is imperative to include the material modeling capabilities into circuit simulators so as to accurately predict the structural, physical, and electrical performances of the materials used in the interconnect structures.

The new developments in electronic system integration drive the need for 3D packaging, system-in-package (SiP), system-on-chip (SOC), stacked integrated circuit (SIC), and 3D integrated circuit (3D-IC) for a variety of reasons, such as miniaturization, heterogeneous integration, improved circuit performance, and lower power consumption [24]. The vertical stacking of the circuits and the overlapping of the metal layers make it increasingly difficult to use the 2D simulators to model the EM reliability of the circuit in these new technologies.

In summary, reported physics-based 3D EM models have the merits of accurate modeling, but they are unable to reflect the reliability of an entire IC. Reported 2D circuit simulators, on the other hand, are able to model the EM reliability at circuit layout level but are unable to consider many important factors for EM in the interconnects in present-day ICs and ICs of the future. It is hence necessary for a complete 3D circuit model to be developed, where electrical, thermal, structural, and material effects are considered in order to correctly assess the EM reliability of the interconnects in an IC. For detail description of the various EM models and their evolution, one may refer to [25].

1.3 Organization of the Book

This brief is organized as follows.

An overview of the EM failures in ULSI interconnects and a discussion on the reported EM models and simulators are presented in this chapter.

Using a simple inverter circuit as an example, the construction of a 3D finite element circuit model from its 2D IC layout, the performance of the transient electro–thermo–structural simulations using both Cadence and ANSYS, and the discussions on the results obtained are presented in Chap. 2. The effects of barrier layer thicknesses and dielectric materials are analyzed as well.

A comparison of the EM performance in a 3D finite element circuit structure and a standard line-via test structure is performed in Chap. 3. The output stage of a class AB amplifier is used as an example. The effects of interconnect structures on the EM reliability of the circuit are also studied.

As an extension to the two-metal layer circuit models in Chaps. 2 and 3, a six-metal layer realistic circuit model with the inclusion of the intra-block interconnects is constructed in Chap. 4. A RF low-noise amplifier (LNA) is used as an example. Modifications to the circuit layout and process conditions are performed based on the EM weak spots identified.

This brief ends with a conclusion and recommended future work in Chap. 5.

1.4 Summary

EM is the main failure mechanism of the interconnects at advanced IC technology nodes. Some fundamentals on EM were introduced, and a discussion on the physics-based EM models at test structure level and the current density–based EM simulators at circuit layout level was presented.

With the inclusion of the factors due to electrical, thermal, and thermo-mechanical stress, the physics-based 3D EM models were able to accurately predict the EM process in the interconnects. However, the complexity of the models limited their application at circuit layout level. The "localized" modeling might not always predict the correct EM failure sites of an IC, especially under the circuit operation condition. On the other hand, the 2D circuit simulators were able to look at the entire circuit as a whole, but they consider current density as the sole driving force for EM, which renders them no longer valid at submicron level. The limitation in dimensionality of the 2D simulators increased the difficulty in including the structural and material effects. Therefore, they were unable to provide an accurate result as the 3D models.

There is a need to combine the benefits of the two and extended the multiple driving forces–based 3D EM model to circuit layout level. The development of the complete 3D circuit modeling is necessary.

References

1. Hu CK, Gignac L, Rosenberg R (2002) Reduced electromigration of Cu wires by surface coating. Appl Phys Lett 81(10):1782–1784
2. Srinivasan J, Adve SV, Bose P, Rivers JA (2004) The impact of technology scaling on lifetime reliability. In: International conference on dependable systems and networks, p 177–186
3. Li BZ, Sullivan TD, Lee TC, Badami D (2004) Reliability challenges for copper interconnects. Microelectron Reliab 44(3):365–380
4. Lloyd JR (1999) Electromigration in integrated circuit conductors. J Appl Phys 32(17):R109–118
5. Understand and avoid electromigration (EM) and IR-drop in custom IP blocks. http://www.synopsys.com/Tools/Verification/CapsuleModule/CustomSim-RA-wp.pdf
6. Strong AW, Wu EY, Vollertsen R-P, Sune J, Rosa GL (2009) Reliability wearout mechanisms in advanced CMOS technologies. Wiley, London
7. Tan CM (2010) Electromigration in ULSI interconnects. World Scientific, New York
8. Meier AV (2006) Electric power systems: a conceptual introduction, Wiley, London
9. Sasagawa K, Nakamura N, Saka M, Abe H (1998) A new approach to calculate atomic flux divergence by electromigration. Trans. ASME J Electron Packag 120:360
10. Sasagawa K, Nakamura N, Saka M, Abe H (1999) A method to predict electromigration failure of metal lines. J Appl Phys 86:6043
11. Rzepka S, Meusel E, Korhonen MA, Li C-Y (1999) 3-D finite element simulator for migration effects due to various driving forces in interconnect lines. In: Stress-induced phenomena in metallization: fifth inter workshop, vol 491: AIP, pp 150–161
12. Dalleau D, Weide-Zaage K (2001) Three-dimensional voids simulation in chip metallization structures: a contribution to reliability evaluation. Microelectron Reliab 41(9–10):1625–1630
13. Sheu BJ, Hsu W-J, Lee BW (1989) An integrated circuit simulator- RELY. IEEE J Solid-State Circuits SC-24:473–477
14. Najm F, Burch R, Yang P, Haij IN (1990) Probabilistic simulation for reliability analysis of CMOS VLSI circuits. IEEE Trans. Comput-Aided Des Integr Circuits Syst 9(4):439–450
15. Frost DF, Poole KF (1989) Reliant: a reliability analysis tool for VLSI interconnects. IEEE J Solid-State Circuits 24(2):458–462
16. Hu CM (1998) BERT: an IC reliability simulator. Microelectron J 23:97–102

17. Complete timing signoff in the nanometer era. http://w2.cadence.com/whitepapers/timing_signoff_wp.pdf
18. Accurately measuring specimen temperature in xenon-arc accelerated weathering instruments. http://atlas-mts.com/technical-information/sunspots/current-issue/
19. Teng C, Cheng Y, Rosenbaum E, Kang S (1997) iTEM: a temperature-dependent electromigration reliability diagnosis tool. IEEE Trans. Comput-Aided Des Integr Circuits Syst 16:882–893
20. Alam SM, Troxel DE, Thompson CV (2003) Layout specific circuit evaluation in three-dimensional integrated circuits. Analog Integr Circ Sig Process 35(2–3):199–206
21. Alam SM, Gan CL, Thompson CV, Troxel DE (2004) Circuit level reliability analysis of Cu interconnects. In: Proceedings 5th international symposium quality electronic design, p 238–243
22. Alam SM, Gan CL, Thompson CV, Troxel DE (2007) Reliability computer-aided design tool for full-chip electromigration analysis and comparison with different interconnect metallizations. Microelectron J 38:4–5
23. Li W, Tan CM, Hou Y (2007) Dynamic simulation of electromigration in polycrystalline interconnect thin film using combined Monte Carlo algorithm and finite element modeling. J Appl Phys 101(10):104314
24. International technology roadmap for semiconductors 2009 chapter for interconnects. http://www.itrs.net/Links/2009ITRS/2009Chapters_2009Tables/2009_Interconnect.pdf
25. Tan CM, Li W, Gan Z, Hou Y (2011) Applications of finite element methods for reliability studies on ULSI interconnections. Springer series in reliability engineering, 1st edn, pp 72–112

Chapter 2
3D Circuit Model Construction and Simulation

2.1 Introduction

Chapter 1 shows the need to model the EM reliability of ULSI interconnects using 3D model at circuit layout level. In order to perform such modeling, a method to construct a complete 3D circuit model is necessary and this chapter will illustrate the construction and the corresponding transient electro-thermo-structural simulations for the EM reliability assessment of an IC.

Power and ground rails used to be the major concerns for the EM failure of the circuit due to the large, unidirectional currents in these lines. As the technology node goes down to 65 nm and below, the EM effects on signal wires and even in the logic cells themselves become significant [1]. To evaluate the EM reliability of a circuit, we should therefore look at both the power and the signal wires, as well as the global and the local interconnects.

To show a step by step approach, our illustration is carried out on a small circuit with two transistors, and only the transistor region (i.e., a simplified circuit structure that includes only the local interconnects with two metal levels) is considered. The circuit chosen is a simple inverter circuit and it is shown in Fig. 2.1. The extension to the real circuit structure with the inclusion of other circuit components and the global interconnects up to the metal top will be discussed in detail in Chap. 4.

2.2 Layout Extraction and 3D Model Construction

CMOS inverter is a basic building block in digital circuits [2]. Clock drivers made of inverter trees are prone to EM failure due to the large current demand on every cycle. The circuit layout of the inverter circuit shown in Fig. 2.1 is drawn with Cadence using 0.18 μm technology based on the standard cell design of Global Foundry, as shown in Fig. 2.2. The sizes of the PMOS and NMOS are kept small so as to reduce the size of the 3D model that is going to be built.

C. M. Tan and F. He, *Electromigration Modeling at Circuit Layout Level*,
SpringerBriefs in Reliability, DOI: 10.1007/978-981-4451-21-5_2, © The Author(s) 2013

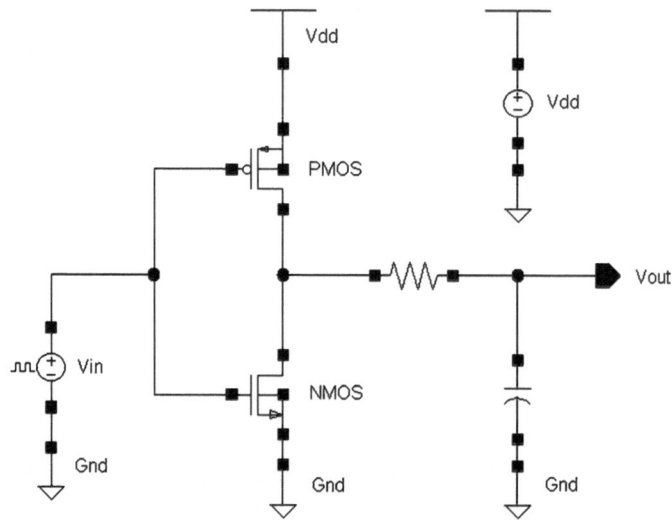

Fig. 2.1 Schematic of a simple inverter circuit

The 0.18 μm technology is chosen because the technology file is readily available and it will also be easier for the future experimental verification as the process technology is mature. As such, unnecessary unknown reliability issues can be avoided during the experimentation. The circuit designs using other technologies can also be used for the 3D model construction with the same method introduced here.

The Cadence GDSII file is first converted, layer by layer, into an ANSYS compatible file using third party software (e.g., LinkCAD). The thicknesses of each layer are specified during the conversion. Since our focus is on the EM reliability of interconnects, the transistors are assumed to be reliable. When the circuit is operating, the current flowing in the interconnects causes Joule heating and affects the temperature of the circuit, especially when the current spikes occur during switching. The focus of attention, the interconnect layers, together with the heat sources which are the transistor diffusion regions, are extracted and imported into ANSYS WORKBENCH. The converted layers are then combined and a 3D model is built [3]. As the layout tool provides only the basic interconnection structures, Ta barrier layer of 25 nm and SiN cap layer of 50 nm [4] are added around the interconnects, as shown in Fig. 2.3 [5, 6].

After importing the essential structures, additional layers that are present in the chip, such as the Si substrate, inter-level dielectric (ILD), passivation, as well as the packaging materials such as the plastic encapsulation, die attach and die pad (e.g., metal plate) as shown in Fig. 2.4 are added to the model, so as to represent a realistic chip condition. This will also enable us to perform chip-package interaction study, if needed. However, such interaction is outside the scope of this work.

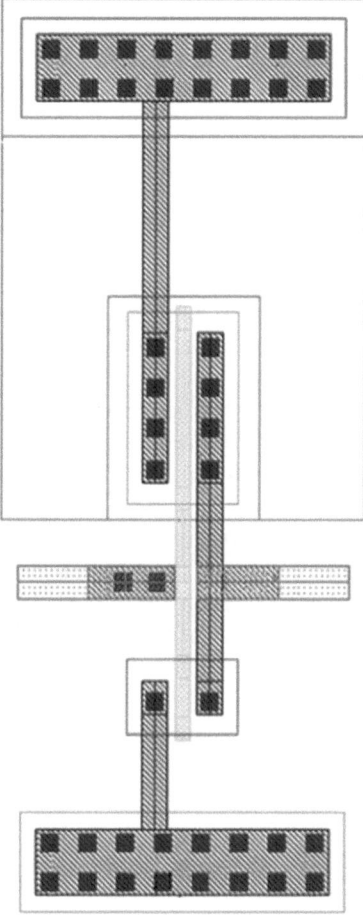

Fig. 2.2 Layout of a simple inverter circuit

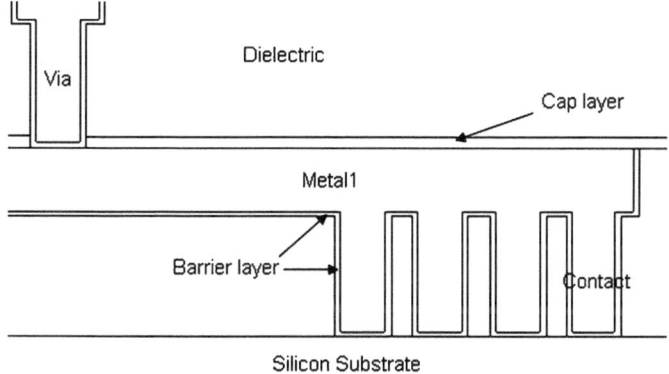

Fig. 2.3 Side view of the interconnects under study showing the barrier and cap layer (Reprinted from [6], with permission from Elsevier)

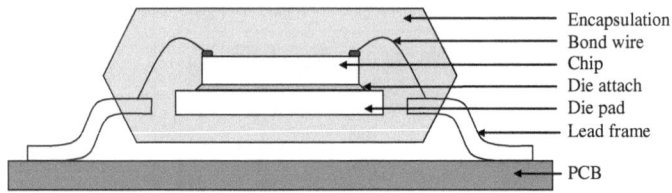

Fig. 2.4 Illustration of the plastic encapsulation used in a microelectronic device

To reduce the area effect on the circuit temperature, the substrate and the encapsulation are drawn to be much larger than the size of the inverter circuit. The complete 3D model with a size of 1,000 × 1,000 × 400 μm (i.e., the size of a die) is shown in Fig. 2.5. The inverter circuit is enclosed at the center just above the Si substrate, as indicated by the rectangular box in Fig. 2.5. A zoom in view of the inverter circuit is shown in Fig. 2.6. PS, PD, NS, and ND in Fig. 2.6 denote the PMOS source, PMOS drain, NMOS source, and NMOS drain respectively.

The geometrical parameter, the properties of the materials used, and the variation of thermal conductivity with temperature for the materials of interest are listed in Tables 2.1, 2.2, and 2.3 respectively [7–11]. The thermal conductivity of the encapsulation is assumed to be temperature independent in this case as the test temperature is below its glass transition temperature which is usually taken as 260 °C [12].

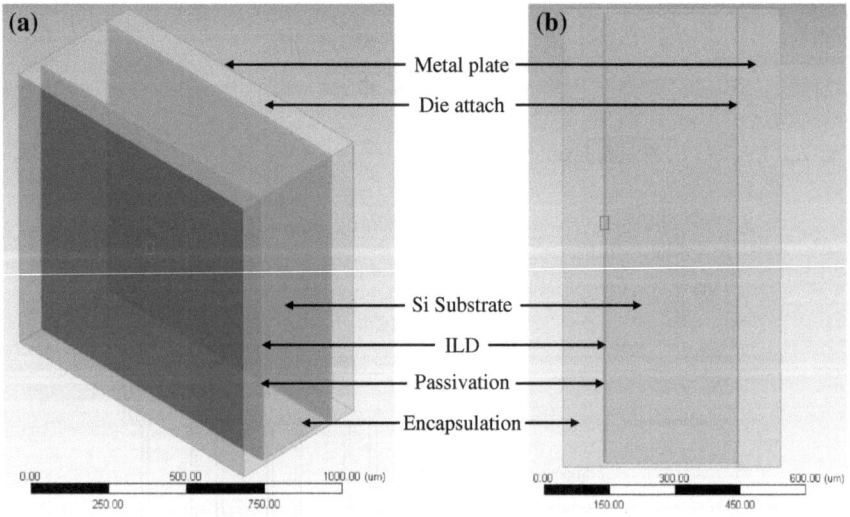

Fig. 2.5 a Isometric and **b** side view of the 3D model. The chip is modeled in a package with encapsulation and die attached to the lead frame represented by the metal plate shown in the figure. The model is in half transparent view for the ease of showing different layers

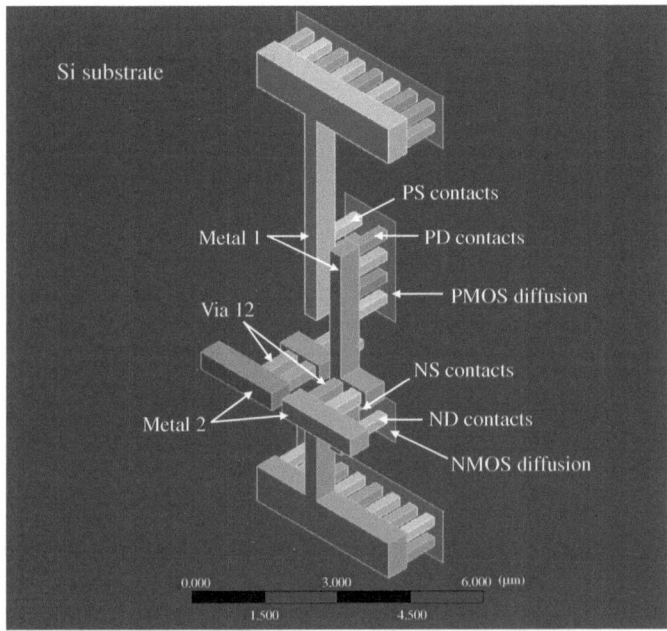

Fig. 2.6 Zoom in isometric view of the inverter circuit after removing the covered layers

Table 2.1 Geometry parameters of the model [8]

Feature	Size
Contact size	0.22 μm
Contact space	0.305 μm
Contact height	0.85 μm
Metal overlap contact	0.06 μm
Metal 1/Metal 2 width	0.34/0.42 μm
Metal 1/Metal 2 height	0.53 μm
Via 12 size	0.26 μm
Via 12 space	0.26 μm
Via 12 height	0.85 μm
PMOS diffusion region dimension (width × length)	2.46 × 1.46 μm
NMOS diffusion region dimension (width × length)	0.96 × 1.46 μm
Si substrate thickness	300 μm
Si substrate dimension (width × length)	1,000 × 1,000 μm
ILD thickness	3.10 μm
Passivation thickness	3.25 μm
Die attach thickness	2 μm
Metal plate thickness	100 μm
Encapsulation thickness	400 μm
Encapsulation dimension (width × length)	1,020 × 1,020 μm
Barrier layer thickness	25 nm
Cap layer thickness	50 nm

2.3 Transient Electro-Thermo-Structural Simulations and Atomic Flux Divergence Computation

After the construction of the 3D finite element circuit model, the next step is to perform the EM analysis. Different from the work reported in literature [13–16] which used DC analysis, the simulation of our 3D circuit model is performed under transient condition so as to account for the possible temperature and stress effects due to the change in magnitude and/or direction of the current flow. Coupled thermal-electric and structural-thermal simulations are performed to calculate the AFDs in the interconnects due to different driving forces, and these values are used to evaluate the EM reliability of the interconnects.

2.3.1 Transient Thermal-Electric Analysis

The transient thermal-electric analysis is performed to study the effect of the electrical loads on the temperature and current density distributions. For an accurate simulation, the boundary conditions and the electrical loads should be properly set.

2.3.1.1 Boundary Conditions

The running speed, the cooling method, and the operation modes of the transistors determine the temperature of an IC. The desktop processors usually run at a temperature between 70 and 90 °C [17]. In this work, a worst-case temperature of 90 °C is used to represent the interconnect heating in a chip under operation [18]. Other temperatures can be used depending on the application of the circuits. The metal base plate is treated as the heat sink since its thermal conductivity is more than 100 times better than that of the encapsulation material as can be seen in Table 2.2, and thus it is the major heat dissipation path for the model. A convection heat transfer coefficient h of 20 W/m$^2 \cdot$ °C [19] is used to simulate the 90 °C ambient condition. A constant h is used for simplicity as the convection has minor effect to the model temperature in this work as verified by our simulation where the model temperature is found to remain unchanged when the value of h varies from 10 to 40 W/m$^2 \cdot$ °C. The boundary conditions of the thermal-electric simulation used in the model are shown in Fig. 2.7.

2.3.1.2 Electrical Loads

The instantaneous source and drain currents flowing in the interconnects, together with the voltages across the interconnects, are the causes of the heat up of the

Table 2.2 Material properties used in the model [7, 9, 10]

Materials	Properties						
	Young's Modulus (MPa)	Poisson's Ratio	Density (kg/m³)	Thermal Expansion (1/°C)	Thermal Conductivity (W/m·°C)	Specific Heat (J/kg·°C)	Resistivity (Ohm·m)
Cu	1.10×10^5	0.34	8,300	1.80×10^{-5}	403	385	1.72×10^{-8}
Si	1.30×10^5	0.28	2,330	2.60×10^{-6}	149	700	1.00×10^{17}
Substrate							
As-doped Si							9.90×10^{-3}
B-doped Si							2.81×10^{-2}
Heavily doped Si							1.00×10^{-5}
SiO₂	7.14×10^4	0.16	2,200	6.80×10^{-7}	1.38	1000	1.00×10^{17}
SiN	2.20×10^5	0.27	3,100	3.20×10^{-6}	30	700	1.00×10^{13}
Ta	1.86×10^5	0.34	16,690	6.30×10^{-6}	57.50	140	1.31×10^{-7}
Polyimide	3,100	0.33	1,430	5.00×10^{-5}	0.18	1100	1.00×10^{15}
96 % Alumina	276	0.25	3,965	7.10×10^{-6}	20.90	779	1.00×10^{13}

Table 2.3 Temperature dependency of thermal conductivity for some materials [11]

Materials	Thermal conductivity (W/m · °C)					
	0 °C	27 °C	127 °C	227 °C	327 °C	427 °C
Cu	403	395	390	380	370	365
Si Substrate	149	148	99	76	62	51
SiO$_2$	1.38	1.38	1.51	1.61	1.75	1.92

Reprinted from He and Tan [5], with permission from Elsevier

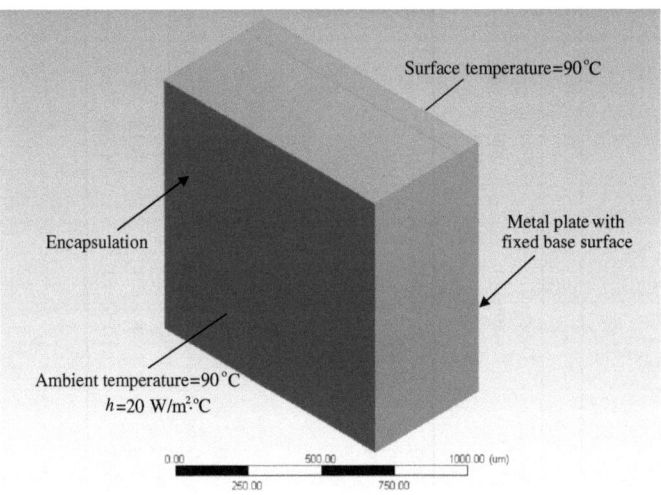

Fig. 2.7 Boundary conditions of the thermal-electric simulation

interconnects. These currents and voltages are treated as the electrical loads in the ANSYS thermal-electric simulation, and their values are determined from the circuit design tool, Cadence.

The power consumed in the inverter circuit includes the dynamic power due to the charging and discharging of the capacitors, and the short circuit power due to the direct current flow from the power source to the ground during switching when both transistors are on [20]. The diode leakage power is assumed to be negligible and is ignored in this simulation. To simulate the power consumption in the inverter, both the charging and discharging currents and the short circuit current should be considered.

When PMOS is on, the path of the current flow is from the V_{dd} line to the source of PMOS, and then from the drain of PMOS to the output line. For the NMOS-on condition, the current flows from the output line through the NMOS to the ground. The current flow paths and the name of the four current sources are listed below:

1. the current flow from the V_{dd} line to the source of PMOS: I_{ps},
2. the current flow from the drain of PMOS to the output line: I_{pd},
3. the current flow from the output line to the drain of NMOS: I_{nd},

4. the current flow from the source of NMOS to the ground: I_{ns}.

The above charging and discharging currents are represented by the gradual increase or decrease of the current waveforms. The short circuit currents are represented by the current spikes during switching and they quickly drop to zero after switching. The currents in the gates of the two transistors are assumed to be negligible.

The root-mean-square (RMS) current density is usually used in the EM simulations performed under DC or low frequency pulsed DC condition. The study in [21] showed that, the steady-state simulation using the RMS current density of 1.5 MA/cm^2 gives a small temperature difference of around 4 % as compared with the transient simulation using an unipolar pulsed DC load at 1 MHz. With a pulsed DC load, EM occurs during the switch-on (i.e., DC) period and nothing or stress relaxation may occur during the switch-off (i.e., no current flow) period [22]. At low frequency of no more than 1 MHz, the metal line can respond thermally to the pulses and thus has a higher temperature during the switch-on period than during the switch-off period. Relaxation of the EM induced stress does not occur during the switch-off period as the temperature of the metal line is low. Therefore the RMS values can be used in the low frequency regime. However, at higher frequencies, the switching speed is faster than the thermal response time of a typical thin-film conductor (i.e., a few μs), stress relaxation occurs as the metallization cannot respond thermally to the pulses [23]. The stress relaxation during the switch-off period results in an enhanced EM lifetime and this effect is overlooked if the RMS value is used. Under an AC operation condition, besides the stress relaxation effect at high frequency, the bi-directional current (e.g., at the output line of the inverter) further affect the stress state in the metal line. It is found in [24] that the EM damage by the positive (forward) current stress can be partially healed by the following negative (reverse) current stress. Therefore it is not suitable to use the RMS values under the AC condition with a frequency higher than 1 MHz, and the instantaneous electrical loads under the AC operation condition are used in our simulation.

The voltage and current pulses at selected circuit nodes (i.e., the source and drain of the two transistors, the V_{dd} and ground lines, the input and output lines) are extracted from Cadence, including both the charging and discharging currents and the current spikes during switching. These current and voltage values are used as the inputs to the 3D circuit model in ANSYS.

In the delay analysis of the CMOS inverter driving a RC load in [25], the load resistance varies from 10 to 1 kΩ, and the load capacitance is in the range of 0.01–1 pF. Here an extreme case of 1 kΩ and 1 pF under an operation frequency of 100 MHz is used as an example. Also, a large load and low frequency are used so that the power consumption of the circuit is larger and the temperature change within one operation cycle is more obvious. Furthermore, the difference between the temperature responses to the rapid current change (e.g. current spike) and the gradual current change (e.g., slow charging or discharging) can be clearer. This non-typical condition is for the validation of the 3D model construction and simulation. Other typical loading and operating frequencies can be used in accordance to the application in a real circuit under the realistic circuit operation condition.

When high frequency AC current flows in a conductor, it tends to redistribute itself within a conductor with the current density being largest near the surface of the conductor, and decreasing at greater depths. The electric current concentrates at the "skin" of the conductor and this phenomenon is called the "skin effect". The depth at which the current density has fallen to 1/e (about 0.37) of its original value at the conductor surface is called the "skin depth" and is calculated as [26],

$$\delta = \sqrt{\frac{2\rho}{\omega\mu}} \tag{2.1}$$

where δ is the skin depth, ω is the angular frequency of current ($2\pi \times$ frequency), and μ is the absolute magnetic permeability of the conductor.

When the skin effect occurs, the effective cross-section of the conductor decrease and its effective resistance increases, especially at higher frequencies where the skin depth is smaller. The increase in the effective resistance of the conductor increases the Joule heating and hence enhances EM. The increased current density at the surface further increases the surface temperature and aggravates the EM process.

At the operating frequency of 100 MHz is used in this case, the skin depth calculated based on Eq. (2.1) is 6.60 μm, and it is larger than the thicknesses of Metal 1 (0.53 μm), Metal 2 (0.53 μm), Via 12 (0.85 μm), and the contacts (0.75 μm), and the skin effect is ignored in this model.

The electrical loads of the inverter circuit within one load cycle (i.e., one period of 10 ns) under the condition specified above are shown in Fig. 2.8. We can see that the PMOS source and drain current waveforms show the same pattern but in opposite direction as expected. The same trend is observed for the NMOS current waveforms. The output voltage does not switch instantaneously with the input voltage due to the presence of the RC load. The threshold voltage V_t of the PMOS and NMOS are -0.495 and 0.482 V respectively, which are obtained from our calculation using Cadence.

As it is not possible to include all the thousands of data points in Fig. 2.8 into the 3D circuit simulation, the data set should be simplified. As can be seen in Fig. 2.8, the electrical loads within one load cycle can be divided into four stages: the first switching stage, the first operation/off stage, the second switching stage, and the second operation/off stage. These four stages are called Stage 1–4 respectively.

For I_{ps} and I_{pd}, there is an only-rise/fall current spike at Stage 1, a gradual current change waveform (i.e., transistor in operation, slow charging of the capacitor) at Stage 2, a rise-and-fall current spike at Stage 3, and a constant current waveform (i.e., transistor is off, almost zero current) at Stage 4. Similarly, for I_{ns} and I_{nd}, there is a rise-and-fall current spike at Stage 1, a constant current waveform (i.e., transistor is off, almost zero current) at Stage 2, an only-rise/fall current spike at

Fig. 2.8 Electrical loads of
the inverter circuit within one
load cycle (10 ns) showing
a the PMOS source and drain
currents **b** the NMOS source
and drain currents and **c** the
input and output voltages

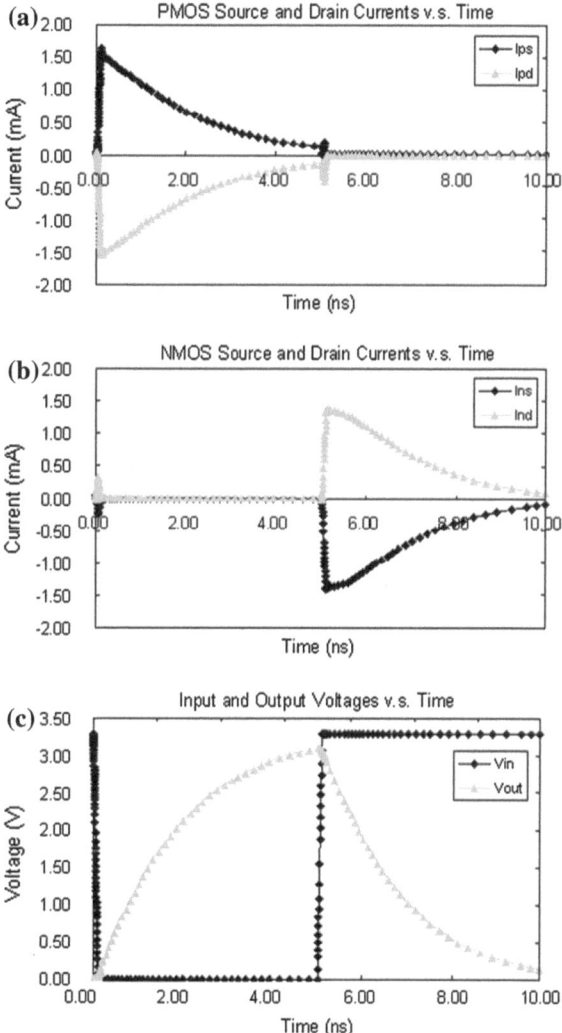

Stage 3, and a gradual current change waveform (i.e., transistor in operation, slow
discharging of the capacitor) at Stage 4. For V_{in}, there is an only-rise/fall voltage
spike at Stage 1 and 3, and a constant voltage waveform at Stage 2 and 4. For V_{out},
there is a gradual voltage increase waveform at Stage 1 and 2, and a gradual voltage
decrease waveform at Stage 3 and 4.

The selection of data points depends on the shape of the current and voltage
waveforms used in the simulation. For example, we should select more data points
at the region with more significant and/or irregular changes. More data points
should be chosen if particular accuracy is required at a specific time interval.

In this case, the gradual current or voltage change waveforms need around
5 data points to represent their shapes. The number of data points selected can

be changed depending on the accuracy requirement and the PC computation capability. Here 5 points are used so that there can be a gradual change in current or voltage at around 1 ns interval. The only-rise/fall current or voltage spikes need 5 data points: 1 at the beginning, 1 at the peak, 3 on the rise/fall slope. The rise-and-fall current spikes need 8 data points: 1 at the beginning, 1 at the peak, 3 on the rise-slope and 3 on the fall-slope. The constant current or voltage waveforms need only 2 data points: 1 at the beginning and 1 at the end. As the current and voltage waveforms occur simultaneously, there are 6 waveforms with different shapes in each stage. When the waveforms within the same stage have different shapes, the number of data points selected is determined by the waveform that needs more points to represent its shape.

Therefore, at Stage 1, the only-rise/fall current spikes for I_{ps} and I_{pd} need 5 data points; the rise-and-fall current spikes for I_{ns} and I_{nd} need 8 data points; the voltage spike for V_{in} need 5 data points; and the gradual voltage increase waveform for V_{out} need 5 data points. Therefore 8 data points are selected at Stage 1. At Stage 2, the gradual current change waveforms for I_{ps} and I_{pd} need 5 data points; the constant current waveforms for I_{ns} and I_{nd} and the constant voltage waveform for V_{in} need 2 data points; and the gradual voltage increase waveform for V_{out} need 5 data points. Therefore 5 data points are selected at Stage 2. At Stage 3, the rise-and-fall current spikes for I_{ps} and I_{pd} need 8 data points; the only-rise/fall current spikes for I_{ns} and I_{nd} need 5 data points; the voltage spike for V_{in} need 5 data points; and the gradual voltage decrease waveform for V_{out} need 5 data points. Therefore 8 data points are selected at Stage 3. At Stage 4, the constant current waveforms for I_{ps} and I_{pd} need 2 data points; the gradual current change waveforms for I_{ns} and I_{nd} need 5 data points; the constant voltage waveform for V_{in} need 2 data points; and the gradual voltage decrease waveform for V_{out} need 5 data points. Therefore 5 data points are selected at Stage 4.

The data point selection at Stages 1 and 4 for I_{nd} and Stage 2 and 3 for I_{ps} is shown in Fig. 2.9. The time scale at Stage 1 and 3 (i.e., the current spike stages) in Fig. 2.9 are exaggerated for better illustration purpose. The data point selection for all the current and voltage waveforms at the four stages is summarized in Table 2.4.

In total, 26 data points are selected to present the shapes of the current and voltage waveforms in Fig. 2.8. The simplified current and voltage waveforms within one load cycle are shown in Fig. 2.10. Each data point corresponds to one load step in the ANSYS simulation. A do-loop is used to repeat the 26 load steps for multiple cycles to represent a continuous circuit operation.

2.3.1.3 Application of the Electrical Loads

The locations for applying the current and voltage loads in the 3D circuit model correspond to the data extraction nodes in Cadence, and they are shown in Fig. 2.11. The direction of the current flow is represented by the + and − sign in the simulation.

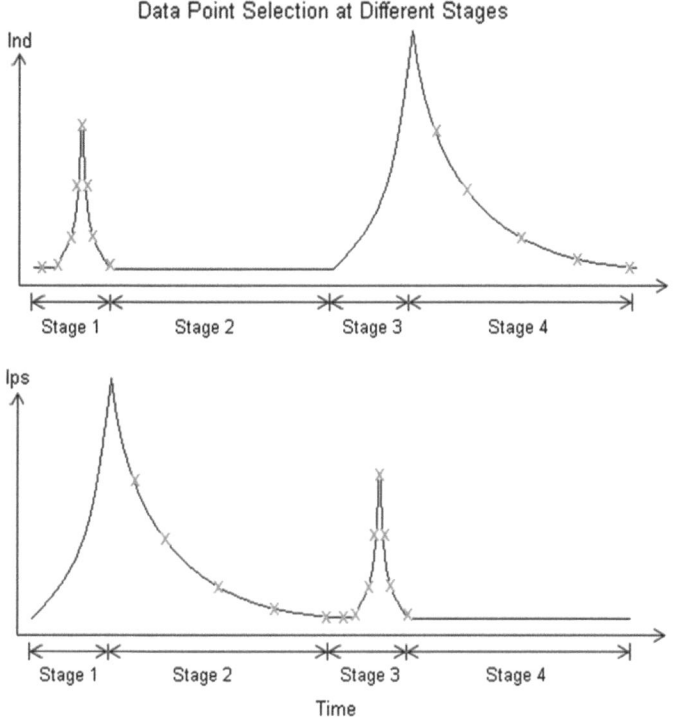

Fig. 2.9 Data point selection at different stages (I_{nd} and I_{ps}), not to scale

Table 2.4 Current and voltage waveforms and the data point selection at different stages. The number in the bracket is the data points selected at this stage for this waveform

Stage	Current waveforms		Voltage waveforms		Data points
	I_{ps} and I_{pd}	I_{ns} and I_{nd}	V_{in}	V_{out}	
1	Only-rise/fall spike (5)	Rise-and-fall spike (8)	Only-fall spike (5)	Gradual increase (5)	8
2	Gradual change (5)	Constant (2)	Constant (2)	Gradual increase (5)	5
3	Rise-and-fall spike (8)	Only-rise/fall spike (5)	Only-rise spike (5)	Gradual decrease (5)	8
4	Constant (2)	Gradual change (5)	Constant (2)	Gradual decrease (5)	5

After the application of the boundary conditions and the electrical loads, the transient thermal-electric simulation is performed and the results are discussed in Sects. 2.4.1 and 2.4.2.

Fig. 2.10 Simplified
electrical loads of the inverter
circuit within one load cycle
(10 ns) showing **a** the PMOS
source and drain currents
b the NMOS source and drain
currents and **c** the input and
output voltages

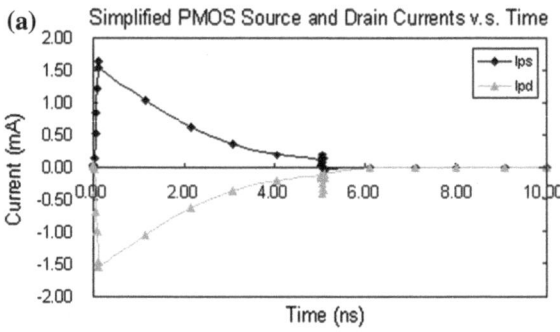

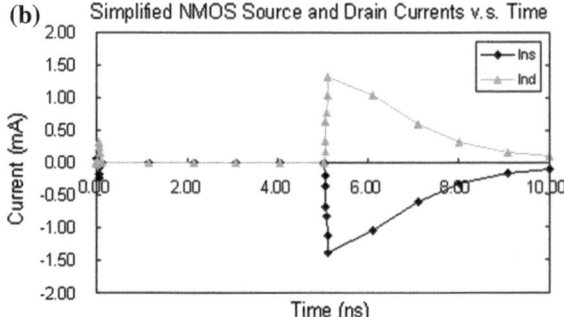

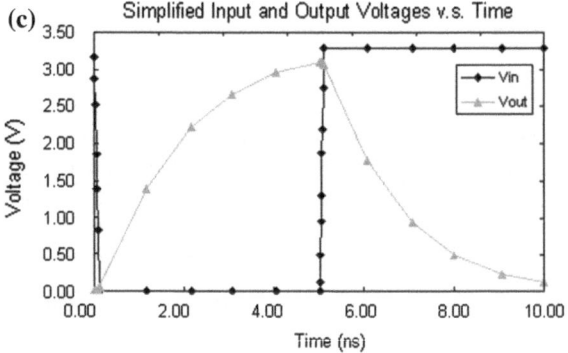

2.3.2 Transient Structural-Thermal Analysis

Besides the intrinsic stress that is already present at the beginning of the simulation, when the current and voltage loads are applied to the model, there is a difference in the interconnect temperature at different locations, and the thermal expansivity mismatch among the materials in an interconnect structure causes thermally induced thermo-mechanical stress. The transient structural-thermal analysis is needed to study the thermo-mechanical stress distribution in the interconnects.

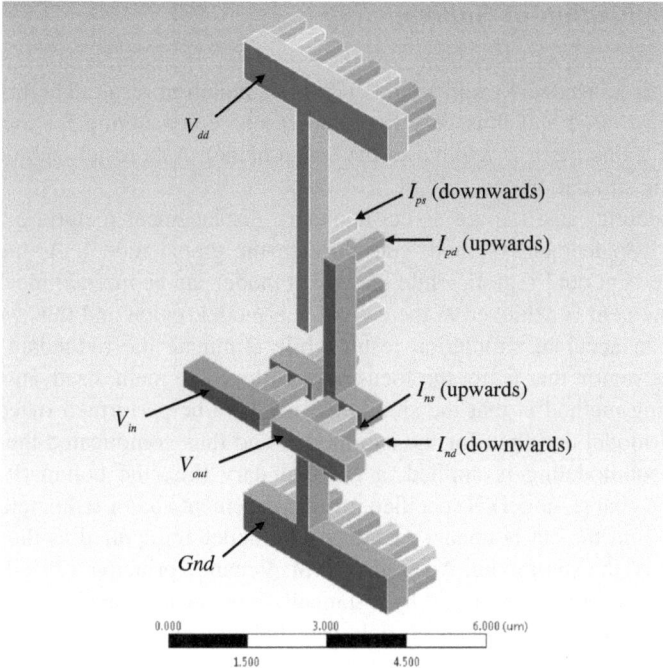

Fig. 2.11 Application of the current and voltage loads to the 3D circuit model in ANSYS, showing the current flow direction during circuit operation. Only the interconnects are shown for clarity

2.3.2.1 Boundary Conditions

All the nodes on the base surface of the model are assumed to be fixed as they are mounted on the circuit board. The stress free temperature (SFT) of the Cu dual-damascene structure is usually between 350 and 400 °C [27, 28], depending on the final annealing temperature of the sample. In this work, a SFT of 350 °C is used.

2.3.2.2 Thermal Loads

The temperatures of all the nodes of the model at each load step are retrieved from the results of the transient thermal-electric simulation. These temperature values, together with the time of their occurrence, are used as the thermal loads in the transient structural-thermal simulation.

After the application of the boundary conditions and the thermal load, the transient structural-thermal simulation is performed and the results are discussed in Sect. 2.4.4.

2.3.3 Application of Submodeling

A fine mesh is needed to achieve an accurate simulation result. The original full model is too large and thus very time and resource consuming for the transient analysis if a fine mesh is applied to the entire model. As a result, submodeling is used in this simulation.

Submodeling, also known as cut-boundary displacement method or specified boundary displacement method, cuts through the global model. A fine mesh is done at the "cut out" region, while the global model can be meshed much coarser. Mesh refinement is achieved at the cut out sub-model region and thus we are able to obtain an accurate simulation result while eliminate the redundant effort on solving the region that is not the focus of attention. The main disadvantage of the submodeling method is that the simulation needs to be performed twice, first for the global model and then for the sub-model, and thus complicated the process.

When submodeling is applied, a cut boundary (i.e., the boundary that cuts through the coarse model) is specified, the displacement and/or temperature results extracted from the cut boundary of the coarse model are applied as the boundary conditions of the sub-model. According to St. Venant's principle [29], if an actual distribution of forces is replaced by a statically equivalent system, the distribution of the stress and strain is changed only near the regions of load application. This implies that, acceptable results can be obtained in the sub-model if the cut boundaries of the sub-model are adequately far away from the stress concentration, in view of the fact that the stress concentration effects are localized around the concentration sites [30].

A simple steady-state simulation is carried out on the full model to determine where the cut boundary should be, i.e., the size of the sub-model. With the same setup for the boundary conditions in Sect. 2.3.1.1, the first current spike is used as the input for the simulation. This current spike is used because it is one of the worst-case values which can result in a higher temperature than that using the normal operation current in the circuit. The resultant worst-case temperature is sufficient for determining the cut boundary location.

From Fig. 2.12, it can be seen that the temperature is almost constant at around 5 μm away from the model center in the x direction and 10 μm in the y direction, as indicated by the rectangular box. Hence it is acceptable to place the cut boundary there. A $15 \times 20 \times 12$ μm sub-model centered with the inverter circuit is cut out from the original $1,000 \times 1,000 \times 400$ μm full model as shown in Fig. 2.13.

The complexity of the model limits the application of the hexahedral mesh. Wang et al. [31] showed that the quadratic tetrahedral shaped element can produce as accurate result as the linear hexahedral element, and hence a quadratic tetrahedral mesh using 20-node coupled-field solid element SOLID98 is used in this work. The quadratic tetrahedral mesh is applied to both the full model and the sub-model. A fine mesh with a size of 0.50 μm is used for the interconnects in the full model, and an even finer mesh is used in the sub-model. A transient analysis using the electrical loads in Fig. 2.10 is conducted to determine the mesh size to be used

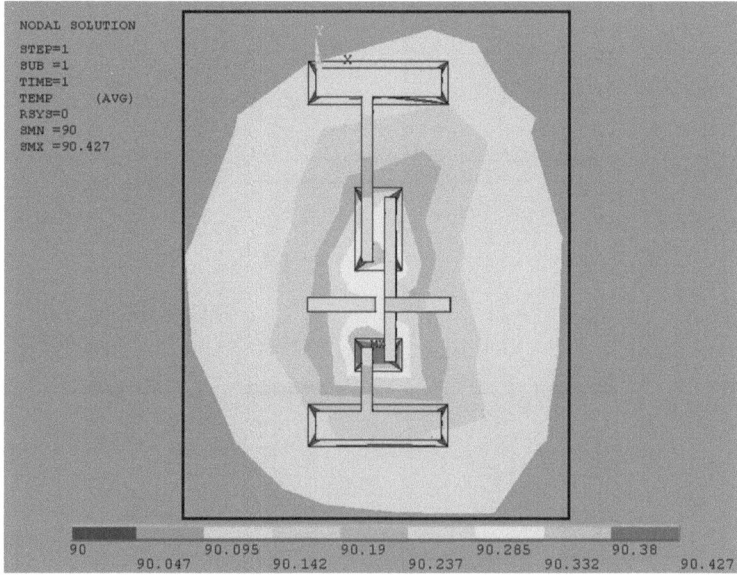

Fig. 2.12 Plane view of the steady-state temperature distribution of the full model using the worst-case current. The *box* indicates the margin where the temperature becomes almost constant

in the sub-model and the simulation results at the end of one load cycle (i.e., 10 ns) are shown in Fig. 2.14.

From Fig. 2.14, it is clear that a fine mesh of 0.20 μm is adequate for the sub-model as the temperature and the stress curves both converge (i.e., the curve becomes flat) at 0.20 μm. Similarly, the barrier layer can be meshed at 0.10 μm without the loss of accuracy. The coarse mesh for the full model and fine mesh for the sub-model are shown in Fig. 2.15.

The outer surfaces of the sub-model are treated as the cut boundary. According to the transient submodeling analysis technique introduced by Wang et al. [32], the simulation results of the full model at each load step are recorded, and the solution of the first load step is retrieved and applied to the cut boundary of the sub-model as the boundary conditions. The sub-model is then solved with the current and the voltage loads and the applied boundary conditions from the first load step. This is repeated for the remaining load steps.

2.3.4 Computation of Atomic Flux Divergences

Finite element analysis using the AFD approach is a common practice in 3D EM modeling. This approach is used as there is a direct relationship between the EM

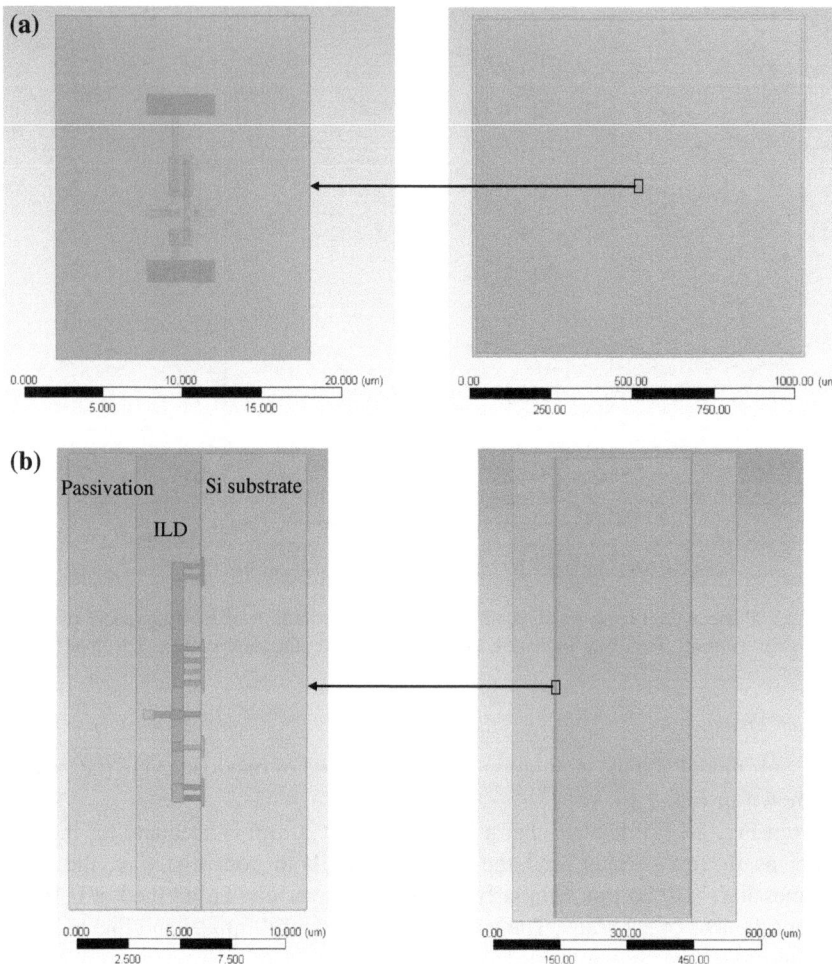

Fig. 2.13 The cut out of the sub-model (*left*) from the full model (*right*), models shown in **a** plane and **b** side views

failure time and the AFD in a small interconnect segment of length Δl for a time period of Δt [33].

$$t_f = \frac{V_c}{\Omega' \Delta l \overline{p_w p_t \nabla \cdot J}} \tag{2.2}$$

where t_f is the time-to-failure, V_c is the critical volume of mass depletion or accumulation required for failure to occur, Ω' is the atomic volume, p_w and p_t are the effective diffusion path width and thickness respectively, and $\nabla \cdot J$ is the total atomic flux divergence.

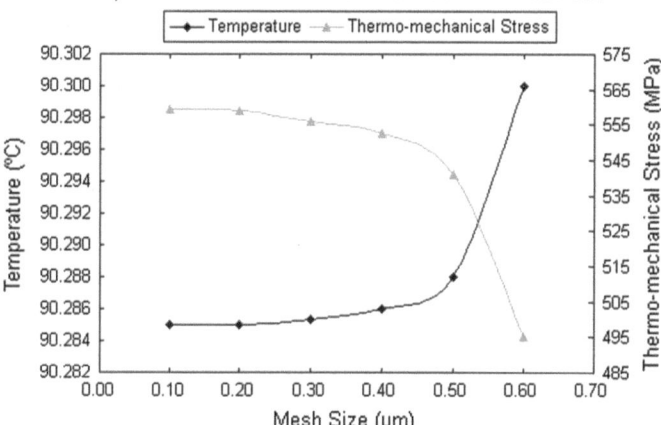

Fig. 2.14 Temperature and thermo-mechanical stress variation with mesh size showing convergence achieves at 0.20 μm

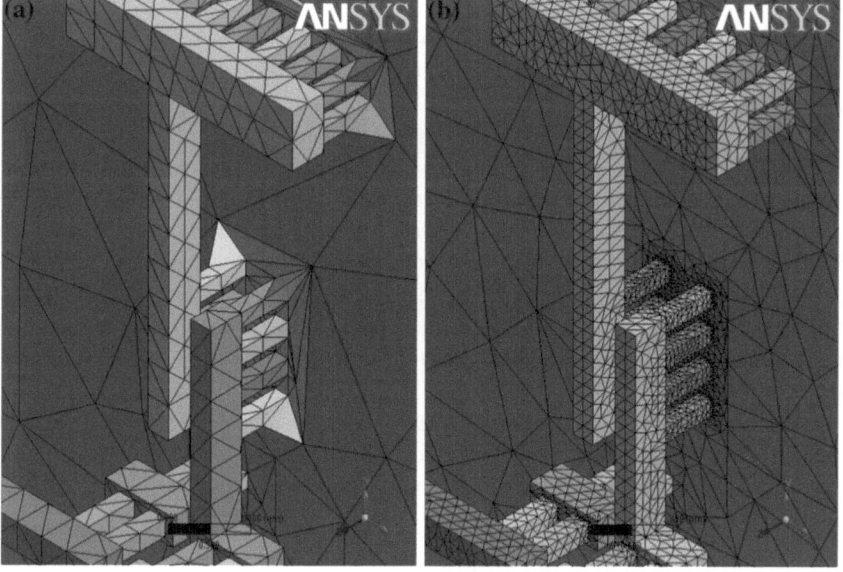

Fig. 2.15 Tetrahedral mesh at the interconnect region showing **a** the coarse mesh for the full model and **b** the fine mesh for the sub-model

From Eq. (2.2), it is obvious that the EM failure time of the interconnects is inversely proportional to the AFD. Void is assumed to nucleate at the location with the highest positive AFD and grows irreversibly. Experimental observation in [34] showed that the void was indeed found at the location of the maximum AFD with excellent agreement, indicating the accuracy of this approach.

The equations used in our model are based on the work by Dalleau et al. [13], and we extend their model from line-via level to circuit layout level. There are three dominant driving forces for the AFD, namely the electron wind force, the temperature gradient induced driving force, and the thermo-mechanical stress gradient induced driving force. The atomic fluxes due to the three driving forces were modeled using Eqs. (2.3)–(2.5) respectively [13],

$$\vec{J}_{EWF} = \frac{N}{k_B T} eZ^* \vec{j} \, \rho D_0 \exp\left(-\frac{E_a}{k_B T}\right) \tag{2.3}$$

$$\vec{J}_T = -\frac{N Q^* D_0}{k_B T^2} \exp\left(-\frac{E_a}{k_B T}\right) \nabla T \tag{2.4}$$

$$\vec{J}_S = \frac{N \Omega' D_0}{k_B T} \exp\left(-\frac{E_a}{k_B T}\right) \nabla \sigma_H \tag{2.5}$$

where J_{EWF}, J_T and J_S are the atomic fluxes due to electron wind force, temperature gradient induced driving force and thermo-mechanical stress gradient induced driving force respectively, N is the atomic concentration, k_B is the Boltzmann constant, T is the temperature of the metal line, eZ^* is the effective charge of the diffusing species, j is the current density, ρ is the electrical resistivity which is dependent on temperature given as $\rho = \rho_0 (1 + \alpha(T - T_0))$, α is the temperature coefficient of resistivity, ρ_0 is the electrical resistivity at temperature T_0, D_0 is the prefactor of the self-diffusion coefficient of the metal line, E_a is the activation energy for self-diffusion of the metal line, Q^* is the heat of transport of the metal line, σ_H is the local hydrostatic stress calculated using the average of the hydrostatic stress values in the interconnect in the x, y, z direction as $\sigma_H = (\sigma_{xx} + \sigma_{yy} + \sigma_{zz})/3$.

After obtaining the atomic fluxes, the divergences due to electron wind force (AFD_EWF), temperature gradient induced driving force (AFD_T), and thermo-mechanical stress gradient induced driving force (AFD_S) were calculated using Eqs. (2.6)–(2.8) respectively [13]:

$$\nabla \cdot \vec{J}_{EWF} = \left(\frac{E_a}{k_B T^2} - \frac{1}{T} + \alpha \frac{\rho_0}{\rho}\right) \cdot \vec{J}_{EWF} \nabla T \tag{2.6}$$

$$\nabla \cdot \vec{J}_T = \left(\frac{E_a}{k_B T^2} - \frac{3}{T} + \alpha \frac{\rho_0}{\rho}\right) \cdot \vec{J}_T \nabla T + \frac{N Q^* D_0}{3 k_B^3 T^3} \vec{j}^2 \rho^2 e^2 \exp\left(-\frac{E_a}{k_B T}\right) \tag{2.7}$$

$$\nabla \cdot \vec{J_S} = \left(\frac{E_a}{k_B T^2} - \frac{1}{T}\right) \cdot \vec{J_S} \nabla T + \frac{2EN\Omega' D_0 \alpha'}{3(1-\nu)k_B T} \exp\left(-\frac{E_a}{k_B T}\right)\left(\frac{1}{T} - \alpha\frac{\rho_0}{\rho}\right)\nabla^2 T$$
$$+ \frac{2EN\Omega' D_0 \alpha'}{3(1-\nu)k_B T} \exp\left(-\frac{E_a}{k_B T}\right)\frac{\vec{j}^2 \rho^2 e^2}{3k_B^2 T}$$

(2.8)

where E, α' and ν are the Young's Modulus, the thermal expansion coefficient, and the Poisson ratio of the metal line respectively.

To compute the AFDs in the interconnects, the thermal-electric simulation is first carried out using the boundary conditions and the electrical loads as shown in Sects. 2.3.1.1 and 2.3.1.2. The nodal temperatures extracted from the thermal-electric simulation are then used as the thermal loads for the structural-thermal simulation. The structural-thermal simulation is performed based on the applied thermal loads and the boundary conditions in Sect. 2.3.2.1. Submodeling technique is used in both simulations. The thermo-mechanical stress gradient of the model is calculated based on the thermo-mechanical stresses extracted from the structural-thermal simulation. The thermo-mechanical stress gradient, together with the current density, temperature, and temperature gradient extracted from the thermal-electric simulation, are used to compute the atomic fluxes and the flux divergences due to electron wind force, temperature gradient induced driving force, and thermo-mechanical stress gradient induced driving force, using Eq. (2.3)–(2.8). The flowchart for the computation of the AFDs in ANSYS is shown in Fig. 2.16. The total AFD in the interconnects is the sum of the three AFDs from their respective driving forces and the results obtained are discussed in Sect. 2.4.5.

The simulations are carried out using an Intel Quad core computer (Q9400@ 2.66 GHz) with 4G RAM. The computation time and result file size of each transient simulation for one load cycle (10 ns, 26 load steps) are listed in Table 2.5 [5].

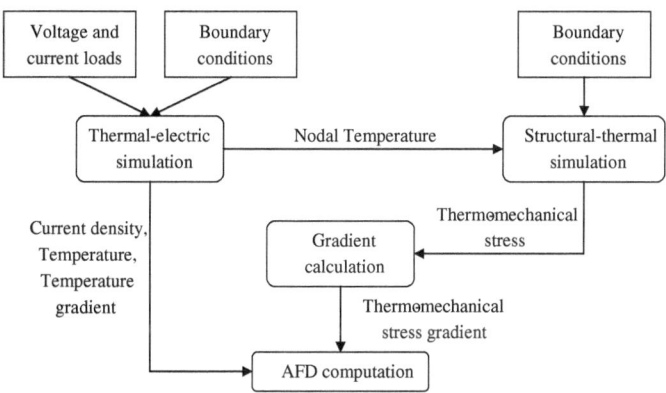

Fig. 2.16 Flowchart for the computation of the atomic flux divergences

Table 2.5 The computation time and result file size of each transient simulation for one load cycle

Simulation	Computation time (min)	Result file size (GB)
Thermal-electric simulation	70	9.81
Structural-thermal simulation	156	9.52
Thermo-mechanical stress calculation & AFD computation	144	6.60
Total	370	25.93

It can be seen from the listed values that the overall simulation takes about 6 h running time and the result file is nearly 26 GB.

Due to the complexity of the problem and the constraint of time and resource, the current simulation is focusing on the void nucleation location instead of the dynamics of void growth. This is justifiable due to the following two reasons. Firstly, recent simulation and experimental results by Li et al. [35] showed that the void nucleation time occupies the largest portion of the EM lifetime, and thus reducing the void nucleation time can effectively enhance the EM lifetime. Secondly, it is always important to identify the weakest site or the damage nucleation site so that one can strengthen these sites for the "design-in" reliability of a device or system.

2.4 Simulation Results and Discussions

2.4.1 Current Density and Temperature Distributions

The current density and temperature distributions of the 3D model at any load step can be viewed after the thermal-electric simulation.

Figures 2.17 and 2.18 show the current density and temperature distributions of the interconnects of the inverter model under study after one and ten load cycles (i.e. 10 and 100 ns) respectively. The figures on the right of Fig. 2.17a are the zoom-in views at the PMOS and NMOS contact regions. For a better illustration, the zoom-in views are drawn using their own contours and SMX and SMN shown on the top right hand side of the zoom-in views are their respective maximum and minimum values.

From Figs. 2.17a and 2.18a, we can see that the current density distributions after 10 ns and 100 ns are similar because the same current pulses are applied to the model. Higher current density (i.e., the red and orange regions in the zoom-in views of Fig. 2.17a) is found at the Via 12/Metal 1 interfaces due to current crowding. The location of the maximum current density depends on the direction of the current flow. For example, the highest current density at the PMOS region is found at the upper corners of the upper most contact at the PMOS source region as

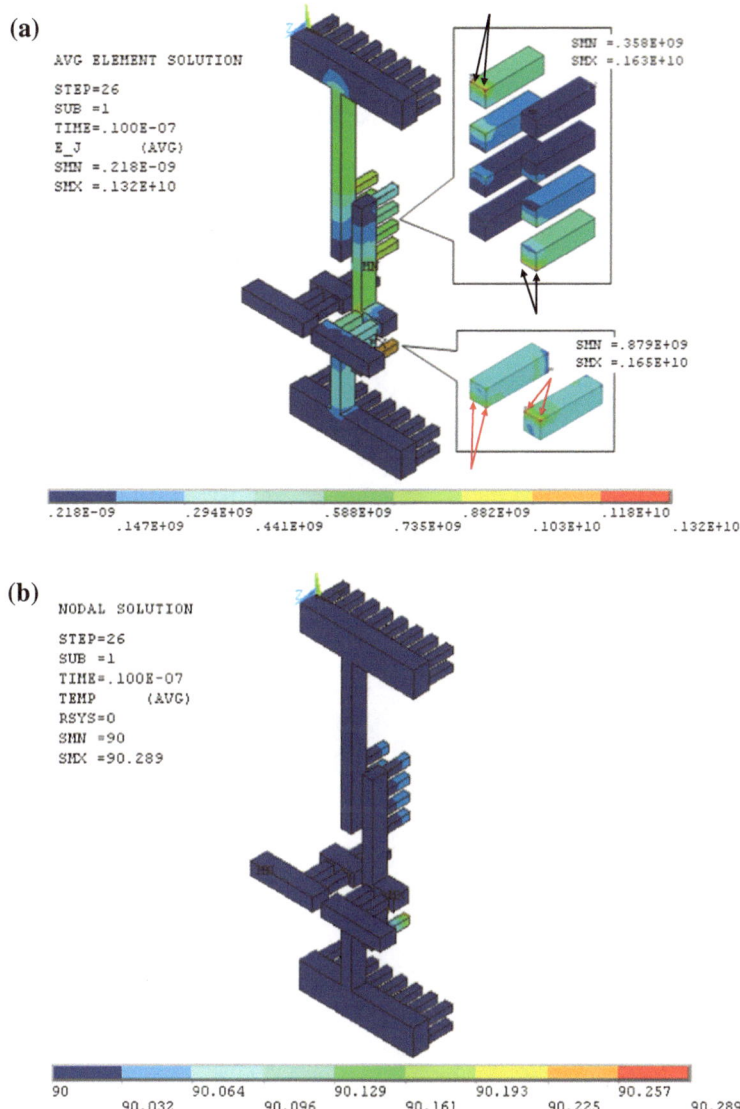

Fig. 2.17 Distributions of **a** current density (unit: pA/μm^2) and **b** temperature (unit: °C) of the interconnects at the end of one load cycle (i.e. 10 ns)

the current is flowing from the V_{dd} line to the PMOS source, while it is found at the lower corners of the lowest contact at the PMOS drain region as the current is flowing from the PMOS drain to the output line, as indicated by the black arrows in Fig. 2.17a. For the NMOS contacts, the maximum current densities are at the lower corners of the source contact and the upper corners of the drain contact, as indicated by the red arrows.

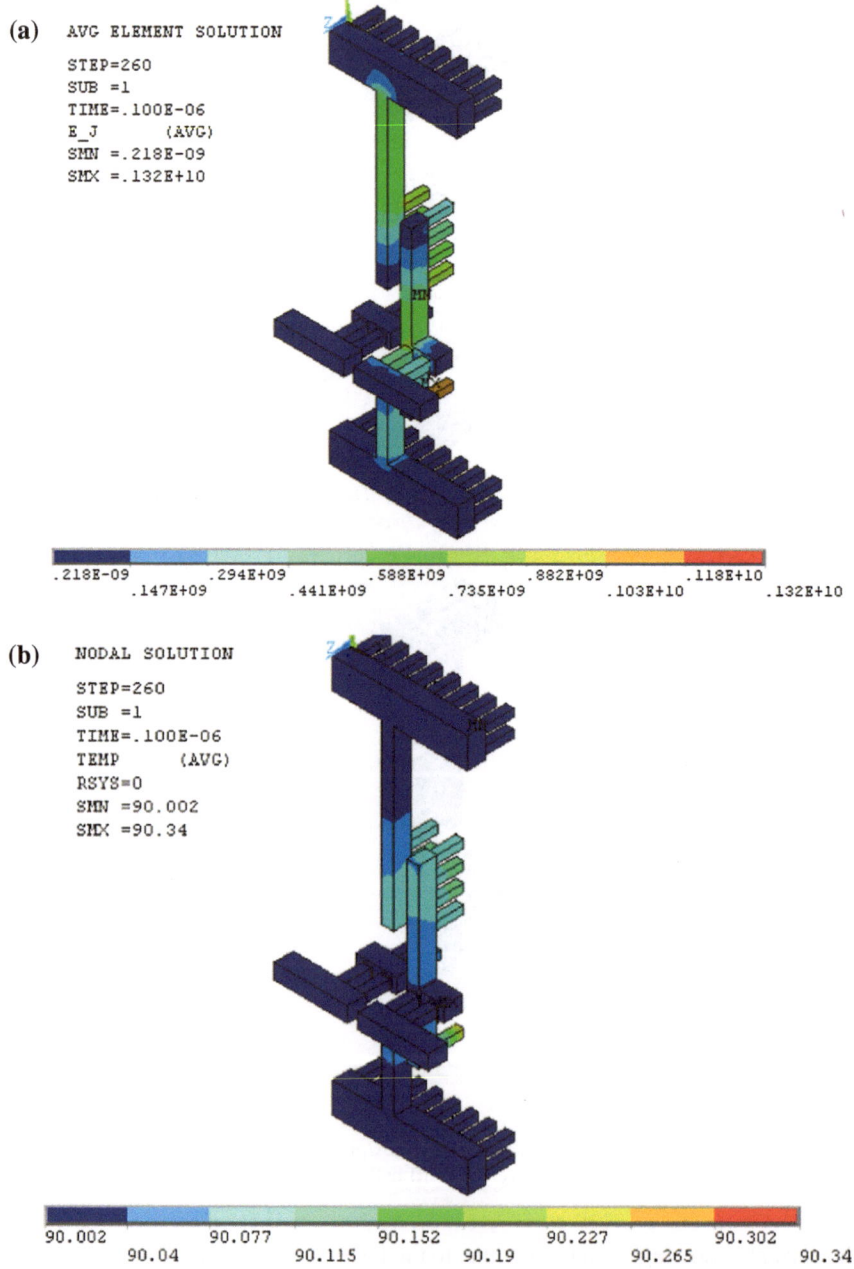

Fig. 2.18 Distributions of **a** current density (unit: pA/μm²) and **b** temperature (unit: °C) of the interconnects at the end of ten load cycles (i.e., 100 ns)

For the temperature distribution, higher temperature is found at the interconnect contact regions as they are nearer to the heat sources, i.e., the diffusion regions. This result is expected as the power consumption of the transistors is large due to the large load the transistors are driving. At the end of one load cycle, NMOS is on and PMOS is off, the temperature at the NMOS contacts at this instant is higher than that at the PMOS contacts (i.e., the light blue region for the PMOS contacts as compared with the green to red regions for the NMOS contacts as shown in Fig. 2.17b). The maximum temperature of the NMOS and PMOS contacts are 90.29 and 90.05 °C respectively. As time progresses, the temperatures at the diffusion regions increase, and this renders an increase in the interconnect temperature, gradually from the contact regions to the upper Metal 1, Via 12 and Metal 2 layers, as showing by the increase in area of the light blue regions in Fig. 2.18b. The location of the maximum temperature remains unchanged.

In the case of the inverter trees with hundreds or thousands of inverters working together, the circuit temperature is expected to be higher due to the interaction of the inverters with each other. The problem is even worse when the loads and the activities of the inverter trees are heavier (e.g., with more fan outs and/or at higher operating frequencies). The increase in temperature increases the total AFD in the interconnects.

2.4.2 Transient Temperature Variation During Circuit Operation

The transient temperature curve at various locations of the model can be plotted from the extracted results of the thermal-electric simulation. Figure 2.19 shows the time variation of the temperature change within one load cycle at the interconnect contact regions.

From Fig. 2.19, it is observed that the temperatures of the PMOS contacts gradually increase from time 0 to 5 ns (i.e., when V_{in} is at 0 V and PMOS is on).

Fig. 2.19 Temperature change with time at the interconnect contact regions within one load cycle

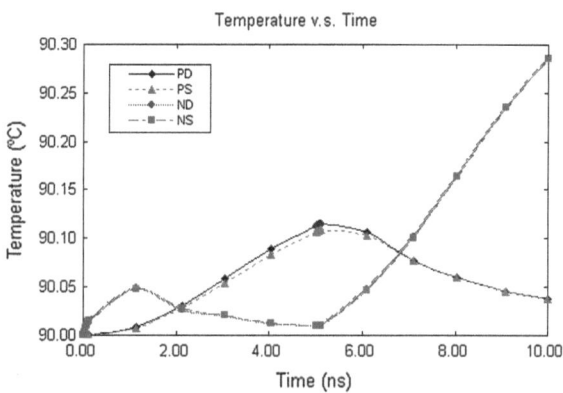

A slight increase in the temperatures of the NMOS contacts at the beginning is due to the current spike occurred during switching. The temperatures of the NMOS contacts drop slowly thereafter, but they do not drop to their initial values.

From 5 to 10 ns, NMOS turns on and PMOS turns off. The temperatures of the PMOS contacts gradually decrease due to heat loss to the surrounding, and they also do not fall back to their initial values. The temperatures of the NMOS contacts start to increase at 5 ns when NMOS begins to operate.

The increase in temperature for the NMOS contacts (around 0.29 °C) when NMOS is on is higher than that for the PMOS contacts (around 0.11 °C) when PMOS is on. This is expected because the size of PMOS in this design is around 2.56 times larger than that of NMOS so as to achieve an electrically symmetric characteristic (i.e., equal rise and fall time at the output), and thus the number of contacts that can be placed at the PMOS region (four pairs) is larger than that at the NMOS region (one pair). With the same current loads, the average current density of the NMOS contacts is higher than that of the PMOS contacts during their respective switch-on period, and thus a higher temperature rise for the NMOS contacts.

For a continuous circuit operation, the temperature variation at the interconnect contact regions shows a zig-zag shape corresponds to the turning on and off of the transistors. A general increasing trend in temperature is observed as shown in Fig. 2.20, which is also in agreement with the simulation result of Alam et al. [36].

Based on Fig. 2.20, the final temperature of the interconnect contact can be obtained through extrapolation using the curve fitting method. The extrapolated curve for the contact at NMOS drain is shown in Fig. 2.21. The final temperature is assumed to be the temperature at time tends to be infinity, which is 106.90 °C in this case.

Fig. 2.20 Temperature change with time at the interconnect contact regions for multiple load cycles. A general increasing trend in temperature is observed

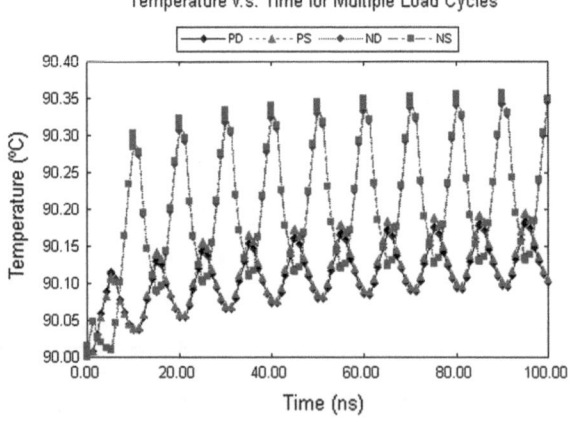

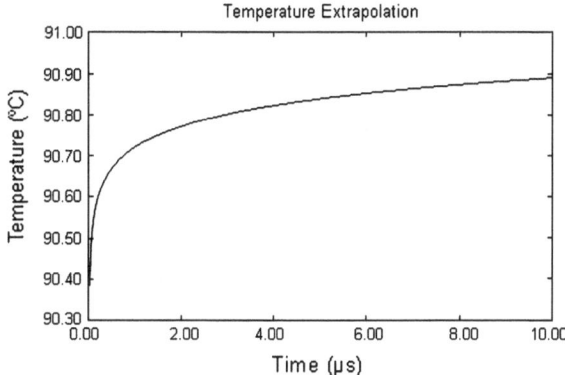

Fig. 2.21 Temperature extrapolation for the contact at NMOS drain

2.4.3 Effect of the Substrate Dimension on the Circuit Temperature

It is expected that the substrate dimension (length × width) can affect the heat dissipation and the circuit temperature. Keeping all other parameters the same, six models with different substrate areas varying from 20 × 20 to 1,000 × 1,000 μm with the substrate thickness fixed at 300 μm are simulated under the same operation condition, and the results are shown in Fig. 2.22.

As the convection of the chip surface to the surrounding has minor impact on the circuit temperature as already explained in Sect. 2.3.1.1, a lower final steady-state temperature is observed for models with larger substrate areas due to their larger conduction areas at the heat sink (i.e., the metal base plate). The rise in the final circuit temperature is small when the substrate dimension decrease from 1,000 × 1,000 to 200 × 200 μm. A very rapid increase in temperature is observed when the dimension of the substrate area is smaller than 200 × 200 μm, showing

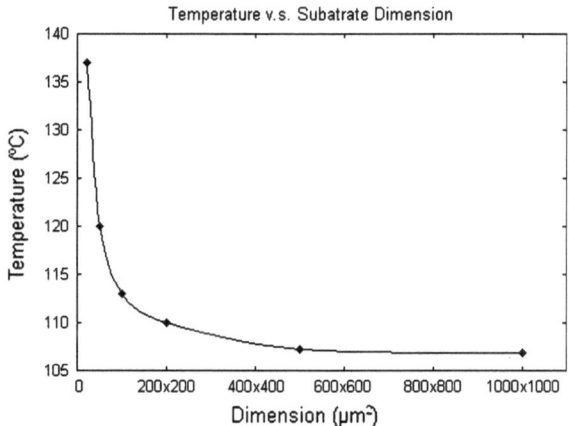

Fig. 2.22 Final circuit temperature of the inverter model with different substrate dimension. The substrate thicknesses of the models are fixed at 300 μm

that the rate of heat loss to the surrounding through the small area is not enough to cater for the rate of increase in temperature of the circuit.

The above results further emphases the necessity of the complete 3D circuit model with the consideration of the size of the surrounding materials as in the actual physical implementation of the circuit.

2.4.4 Transient Thermo-Mechanical Stress Variation During Circuit Operation

The time evolution of the thermo-mechanical stress at any location of the model can be obtained from the results of the transient structural-thermal simulation of the 3D model, and it is as shown in Fig. 2.23. The thermo-mechanical stress distribution of the interconnects at the end of one load cycle (10 ns) is shown in Fig. 2.24.

Intrinsic stress is present at the beginning of the simulation due to the thermal expansivity mismatch between the metallization and its surrounding materials when the metallization is cooled down from the process temperature (i.e. the SFT) to the circuit operation temperature. The total thermo-mechanical stress in the interconnects is the resultant stress of the intrinsic stress and the stress due to the temperature difference in the interconnects caused by the thermal loads. The magnitude of the total thermo-mechanical stress depends on the difference between the SFT of the metallization and the interconnect temperature.

When the interconnect temperature increases and approaches the SFT, the total thermo-mechanical stress in the interconnects decreases and shows an inverse relationship with the interconnect temperature as can be seen in Fig. 2.23. The change in the total thermo-mechanical stress at the end of one load cycle is only around 0.56 MPa and is much smaller than the intrinsic stress which is 559.49 MPa at 90 °C. The change in thermo-mechanical stress is small because the increase in temperature at the end of one load cycle is small (about 0.29 °C).

Fig. 2.23 Comparison of the temperature and thermo-mechanical stress profile of the interconnect contact at NMOS drain

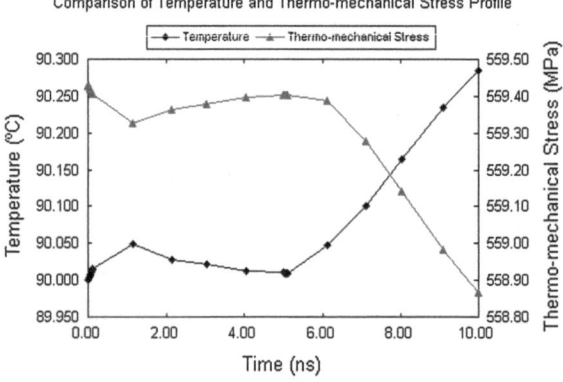

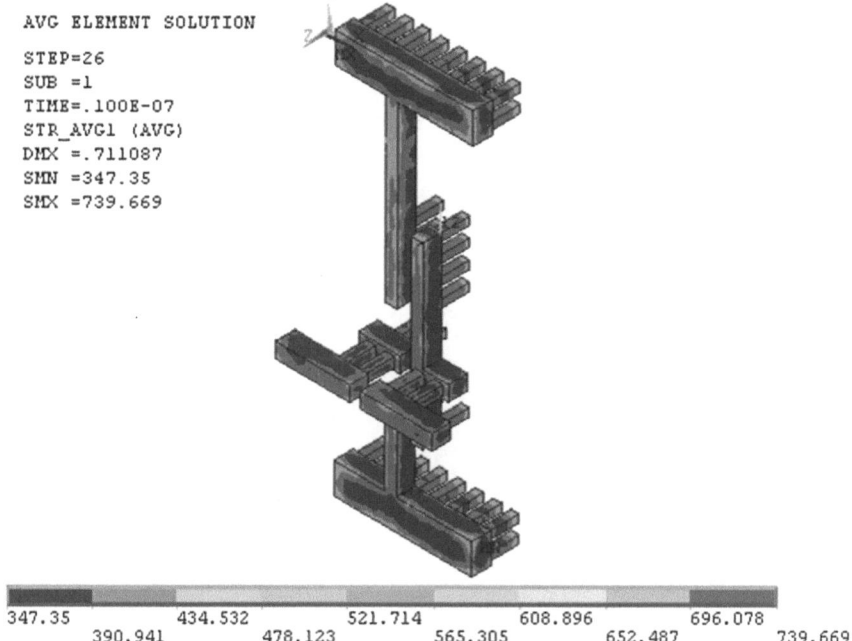

```
AVG ELEMENT SOLUTION
STEP=26
SUB =1
TIME=.100E-07
STR_AVG1 (AVG)
DMX =.711087
SMN =347.35
SMX =739.669
```

347.35   434.532   521.714   608.896   696.078
     390.941   478.123   565.305   652.487   739.669

Fig. 2.24 Thermo-mechanical stress distribution (unit: MPa) of the interconnects at the end of one load cycle

Therefore the thermo-mechanical stress shown in Fig. 2.24 is mainly due to the intrinsic stress in the interconnects, which is more than 50 % higher at the corners (i.e., in contact with other materials) than in the center.

2.4.5 Distributions of the Atomic Flux Divergences

With the current density, temperature, temperature gradient and thermo-mechanical stress gradient obtained from the electro-thermo-structural simulations, the AFDs due to the three driving forces and the total AFD in the interconnects at a specific load step can be obtained.

Figure 2.25 shows the total AFD distribution of the interconnects at the end of one load cycle. The AFD distributions due to electron wind force, temperature gradient induced driving force and thermo-mechanical stress gradient induced driving force are shown in Fig. 2.26a, b, and c respectively.

With recent process improvement on the use of metal cap on Cu interconnects, one can assume a strong Cu/cap layer interface [37], and the void nucleation location (i.e. the EM weak spot) is found at the contact or via bottom as determined by the location of the maximum total AFD [38]. At the end of one load

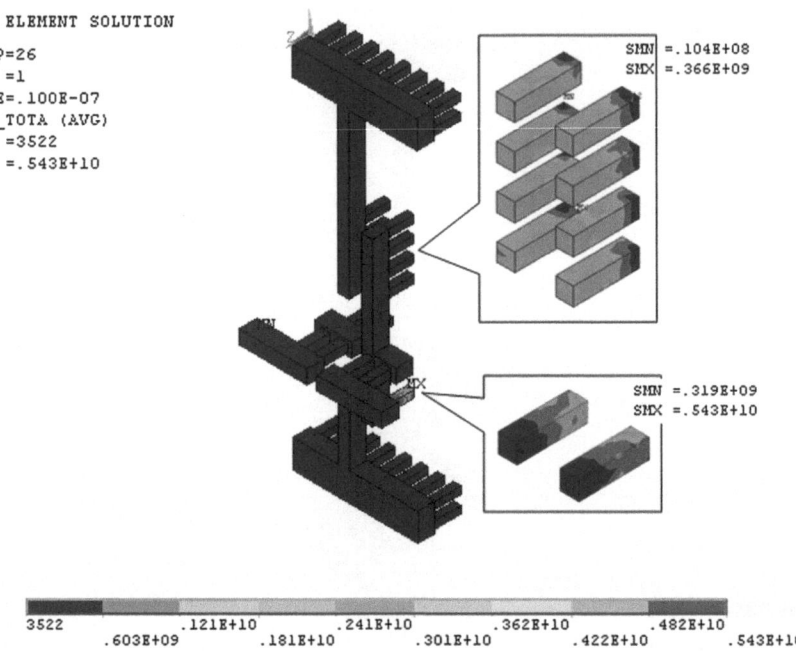

Fig. 2.25 Total AFD distribution (unit: atoms/μm³·s) of the interconnects at the end of one load cycle. The figures on the right are the zoom-in views of the interconnect contacts at the PMOS and NMOS regions

cycle, the maximum AFD location of the inverter model is found at the bottom of the NMOS contacts as indicated by the arrows in Fig. 2.24, which is consistent with the work by Tan et al. [39] and Li et al. [35].

Comparing Figs. 2.25, 2.17a and 2.26a, it is clear that the total AFD distribution is very different from the distributions of the current density and the AFD due to electron wind force (AFD_EWF). At the end of one load cycle, AFD_EWF is 1 and 3 orders of magnitude smaller than the AFD due to temperature gradient induced driving force (AFD_T) and thermo-mechanical stress gradient induced driving force (AFD_S) respectively, as calculated from Eqs. (2.6)–(2.8). In fact, AFD_S has over 99 % contribution to the total AFD and the total AFD distribution of the model actually follows the distribution of AFD_S. In other words, if only current density is considered, the EM weak spot will be different from that if all the driving forces are considered.

From Eq. (2.6), we can see that AFD_EWF depends on the product of the current density and the temperature gradient of the interconnects. For the PMOS contacts, the locations of the maximum current density are the same as the maximum temperature gradient, as indicated by the black arrows in Fig. 2.17a and Fig. 2.27. The locations of the maximum AFD_EWF for the PMOS contacts are found at the upper corners of the upper most contact at the source region and the lower corners of the lowest contact at the drain region, as indicated by the black

Fig. 2.26 AFD distributions of the interconnects due to **a** electron wind force **b** temperature gradient induced driving force and **c** thermo-mechanical stress gradient induced driving force (unit: atoms/μm³·s) at the end of one load cycle. The figures on the *right* are the zoom-in views of the interconnect contacts at the PMOS and NMOS regions

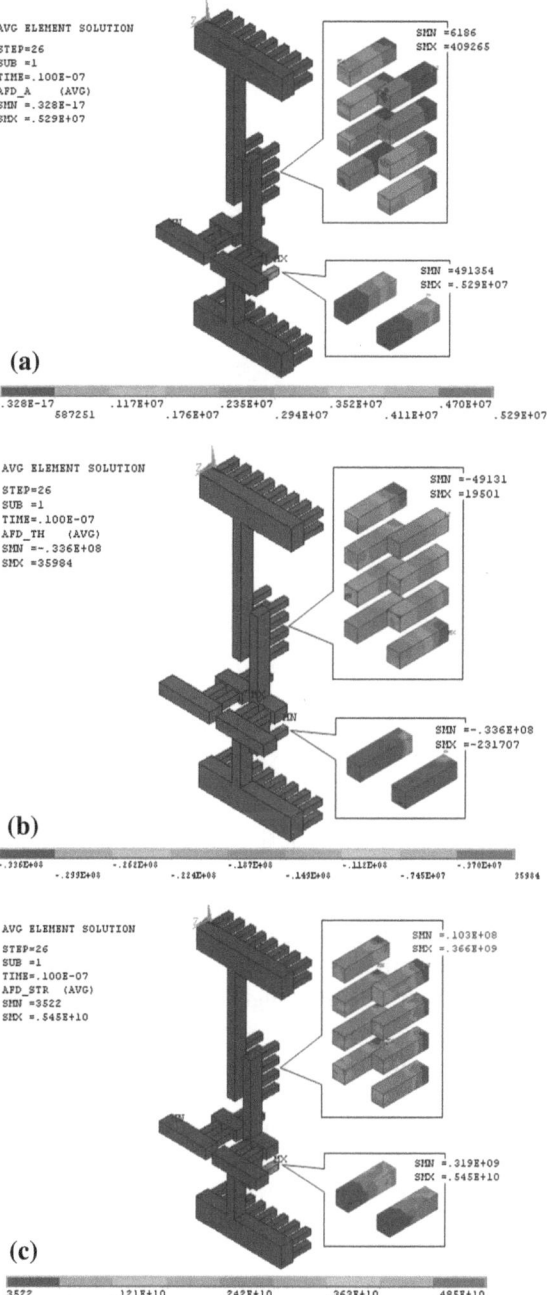

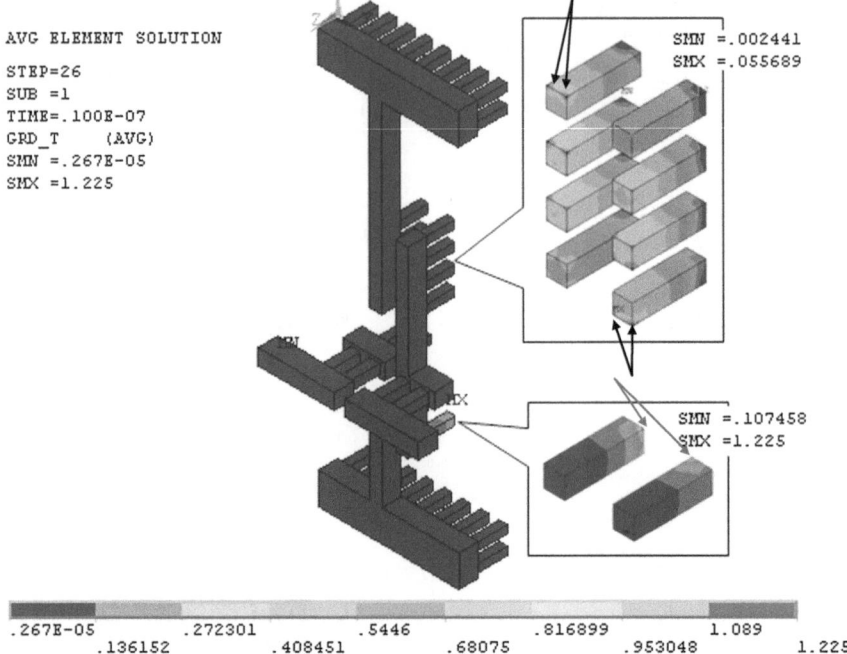

Fig. 2.27 Temperature gradient distribution of the interconnects (unit: °C/μm) at the end of one load cycle. The figures on the *right* are the zoom-in views of the interconnect contacts at the PMOS and NMOS regions

arrows in Fig. 2.26a. For the NMOS contacts, the maximum current densities are at the top of the contacts as indicated by the red arrows in Fig. 2.17a, while the maximum temperature gradients are at the bottom of the contacts as indicated by the red arrows in Fig. 2.27. The temperature gradient at the bottom of the contacts is more than 10 times higher than that at the top while the current density at the top is only less than 2 times higher than that at the bottom. Therefore the locations of the maximum AFD_EWF are found at the bottom of the NMOS contacts, as indicated by the red arrows in Fig. 2.26a.

The maximum AFD due to AFD_T at the interconnect contact region is negative in value as shown in Fig. 2.26b. This is because at the locations with high temperature gradient and comparably small current density (e.g., the red to yellow regions in the zoom-in views of Fig. 2.27 and the blue regions in the zoom-in views of Fig. 2.17a), term 1 in Eq. (2.7) is negative due to its negative coefficient. It dominates over term 2 and results in a negative AFD_T (i.e., the blue to orange regions in the zoom-in views of Fig. 2.26b). At the region with low temperature gradient (e.g., the blue regions in the zoom-in views and blue regions in the global view of Fig. 2.27), term 2 dominates and a positive AFD_T is obtained.

From Eq. (2.8), term 1 dominates over term 2 and term 3 at 90 °C and AFD_S depends on the product of the temperature gradient and the thermo-mechanical

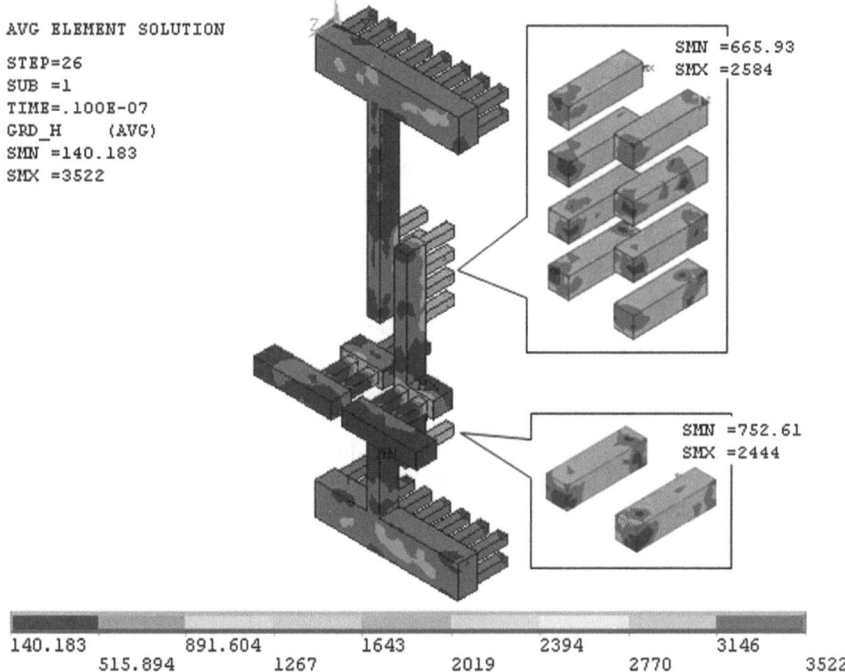

Fig. 2.28 Thermo-mechanical stress gradient distribution (unit: MPa/μm) of the interconnects at the end of one load cycle. The figures on the *right* are the zoom-in views of the interconnect contacts at the PMOS and NMOS regions

stress gradient. The thermo-mechanical stress gradient distribution of the interconnects is shown in Fig. 2.28. This stress gradient is mainly caused by the intrinsic stress in the interconnects. Similar distributions are found for the PMOS and NMOS contacts as they have similar intrinsic stresses. The values of the stress gradient at the two regions are also similar. At the regions where the temperature gradient are less than 0.06 °C/μm (e.g. PMOS contacts, Metal 1, Via 12 and Metal 2), the AFD_S distribution follows the thermo-mechanical stress gradient distribution in the interconnect. At the regions with comparably large temperature gradient (e.g., the bottom of the NMOS contact), the AFD_S distribution follows the temperature gradient distribution. Therefore, to reduce AFD_S and hence the total AFD of the model, one can reduce the thermo-mechanical stress gradient and/or the temperature gradient. This will be studied in Sect. 4.4.

At the beginning of the circuit operation, the temperature and the resultant thermo-mechanical stress of the model, and hence the temperature gradient and the thermo-mechanical stress gradient are changing with time due to the variation in the electrical and thermal loads. The total AFD of the interconnects also changes with time as a result of the changing temperature gradient and thermo-mechanical stress gradient. However, as time progresses, the circuit temperature and temperature gradient become constant and are found to be the highest at the bottom of the NMOS contacts (i.e., the same locations as in Figs. 2.17b and 2.27 but with a higher value). Therefore we can conclude that the maximum total AFD location at the end of one load cycle is the final maximum total AFD location of the model.

In summary, the simulation results in this section show the transient temperature response to the changes in electrical loads and the transient thermo-mechanical stress response to the changes in circuit temperature. This demonstrates the ability of the proposed 3D model for performing the transient temperature and stress analysis at circuit layout level. The distributions of the AFDs are discussed based on the current density, temperature gradient, and thermo-mechanical stress gradient distributions of the interconnects. The void nucleation location and hence the EM weak spot is found at the location with the highest total AFD, which is at the bottom of the NMOS contact for the inverter model under study. It is also found that the electron wind force is no longer the sole driving force for EM as the value and the distribution of the total AFD of the 3D circuit model are determined by AFD_S. To reduce AFD_S and hence the total AFD, the thermo-mechanical stress gradient and/or the temperature gradient of the circuit should be reduced.

2.5 Effects of Barrier Thickness and Low-κ Dielectric on Circuit EM Reliability

This section aims to show the structural and material modeling capability of the 3D circuit model by studying the effects of barrier layer thickness and low-κ dielectric on the EM reliability of the circuit.

Copper has a lower resistivity and a higher melting temperature than Aluminum, and it is replacing the traditional Al-alloy to reduce the RC delay of circuits. However, the high diffusivity of Cu in Si and SiO_2 demands a diffusion barrier layer around the Cu interconnects to prevent the out-diffusion of Cu [40]. A commonly used material for the barrier layer is Ta [41]. As the device feature size goes to 0.18 μm and below, thick barrier layer is found to be undesirable in circuit application due to its high total interconnect resistance [42]. On the other hand, the effectiveness of a diffusion barrier may degrade if its thickness is too thin. Lu et al. [43] showed that thin barrier rendered an early failure and resulted in a shorter EM lifetime due to Cu out-diffusion.

Also, low-κ dielectric is replacing the conventional SiO_2 ILD to further reduce the RC delay so as to meet the demand on the circuit switching speed. However, with a poorer interfacial adhesion and a smaller material elastic modulus, the reliability of the Cu/low-κ structure is a concern [44, 45]. Furthermore, the thermal conductivity of the low-κ dielectric is smaller than that of SiO_2, and this increases the circuit temperature and causes a faster degradation. Using the inverter circuit model in the previous section as an example, the final circuit temperature is 106.90

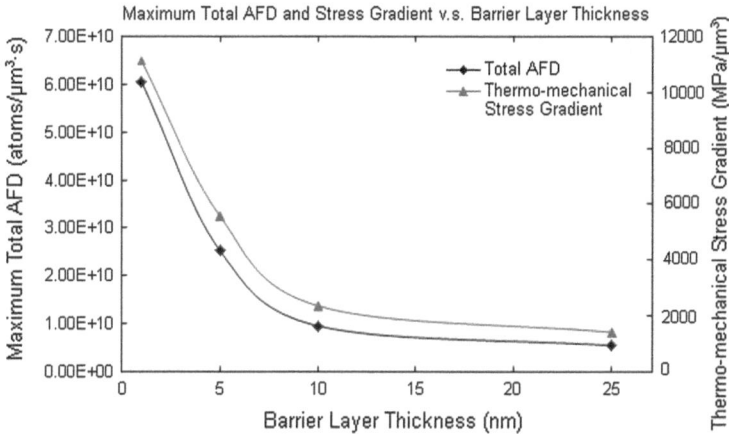

Fig. 2.29 Maximum total AFD and thermo-mechanical stress gradient versus barrier layer thickness for the contact at NMOS drain

and 133.02 °C for the SiO_2 and low-κ (carbon doped oxide) dielectric respectively. The temperature is much higher for the model using the low-κ dielectric due to its poorer thermal conductivity (0.58 W/m·°C) as compared with SiO_2 (1.38 W/m·°C).

To study the effect of barrier layer thickness on the EM reliability of the circuit, four models with the barrier layer thickness varying from 1 to 25 nm are simulated with SiO_2 ILD. The setup for the simulation are the same as in Sects. 2.2 and 2.3 and the simulation results at the end of one load cycle are shown in Fig. 2.29.

From Fig. 2.29, the maximum total AFD is found to increase with decreasing barrier layer thickness, and the rate increases by around 30 times when the barrier layer thickness goes below 7.5 nm, indicating a significant degradation in the EM lifetime with a barrier of thinner than 7.5 nm.

The percentage contribution of the three driving forces AFD_EWF, AFD_T, and AFD_S to the total AFD with different barrier layer thickness at the end of one load cycle is shown in Table 2.6. We can see that AFD_S dominates over the other two driving forces, especially for the models with thinner barriers.

As explained in the previous section, the value of AFD_S at 90 °C is determined by the product of the thermo-mechanical stress gradient and the temperature gradient. With the same electrical loads, the maximum temperature gradients of

Table 2.6 Percentage contribution of the three driving forces to the total AFD

Barrier thickness (nm)	Percentage contribution		
	AFD_EWF	AFD_S	AFD_T
1	0.017	99.855	0.128
5	0.051	99.632	0.317
10	0.075	99.393	0.532
25	0.097	99.284	0.619

the four models with different barrier layer thickness are similar, and it is the difference in the maximum thermo-mechanical stress gradients that causes the difference in the maximum AFD_S and total AFD.

As compared with Cu, Ta has a lower thermal expansion coefficient (6.30×10^{-6} °C^{-1} and 1.80×10^{-5} °C^{-1} respectively) and a higher elastic modulus (1.86×10^5 MPa and 1.10×10^5 MPa respectively), the presence of the Ta barrier introduces a tensile stress in the Cu interconnects in the direction normal to the surface [28, 46]. When the barrier thickness decreases, the confinement effect from the barrier is reduced and this increases the thermo-mechanical stress gradient, as can be seen in in Fig. 2.29. The increase in the thermo-mechanical stress gradient in turn causes an increase in the maximum total AFD. A higher maximum AFD renders a shorter EM lifetime [47], therefore a thinner barrier can cause the reliability degradation of the circuit, and the effect is increasingly significant when the barrier layer thickness goes below 7.5 nm.

Keeping all other conditions the same, various low-κ dielectric, e.g. SiLK, MSQ (methyl silsesquioxane), CDO (carbon doped oxide), and SiCOH are used to replace the traditional SiO$_2$ ILD. The material properties of the low-κ dielectric are listed in Table 2.7, and the values of the maximum total AFD and thermo-mechanical stress gradient of the models with different low-κ dielectric are shown in Fig. 2.30. The barrier layer thicknesses of the models are kept at 25 nm.

As shown in Fig. 2.30, the models using the low-κ dielectric have a higher maximum total AFD than that using the SiO$_2$ ILD due to their higher thermo-mechanical stress gradients. This is because the low-κ materials have lower elastic modulus than SiO$_2$, as shown in Table 2.7. Therefore the confinement effect provided by the elastic property of the low-κ material is lower as proven by the experimental work of Webb et al. [50]. Looking at the high stress gradients due to MSQ and CDO in Fig. 2.30, and that they have much lower elastic modulus than SiO$_2$, one can expect that Cu with MSQ or CDO dielectric suffers significant reduction in the EM lifetime. In addition, as can be seen in Table 2.7, the low-κ materials have higher thermal expansion coefficient (CTE) than SiO$_2$, therefore the difference in CTE between Cu and the low-κ materials is smaller than that between Cu and SiO$_2$. This further reduces the confinement effect and increases the thermo-mechanical stress gradient [28, 46].

Table 2.7 The material properties of the low-κ dielectric as compared with SiO$_2$ [48, 49]

Materials	Properties				
	Young's modulus (MPa)	Poisson's ratio	Density (kg/m^3)	Thermal expansion (1/°C)	Thermal conductivity (W/m·°C)
SiLK	2.50×10^3	0.40	900	6.60×10^{-5}	0.14
MSQ	3.60×10^3	0.25	890	7.30×10^{-6}	0.40
CDO	4.47×10^3	0.30	1,510	1.29×10^{-5}	0.58
SiCOH	1.62×10^4	0.30	910	1.20×10^{-5}	0.40
SiO$_2$	7.14×10^4	0.16	2,200	6.80×10^{-7}	1.38

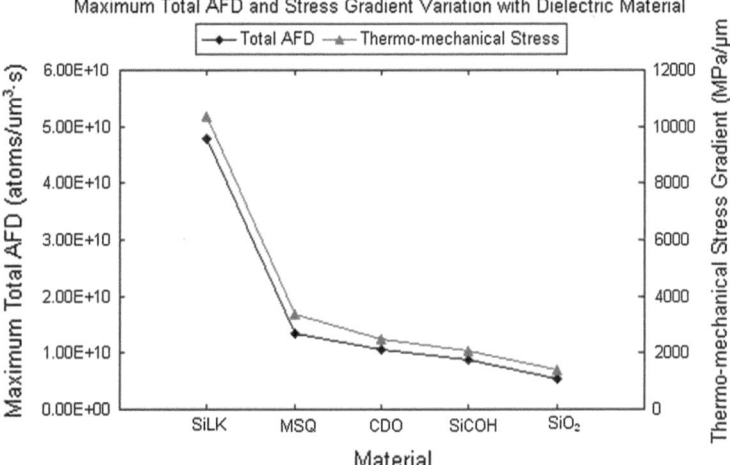

Fig. 2.30 The maximum total AFD and thermo-mechanical stress gradient variation with different dielectric materials for the contact at NMOS drain

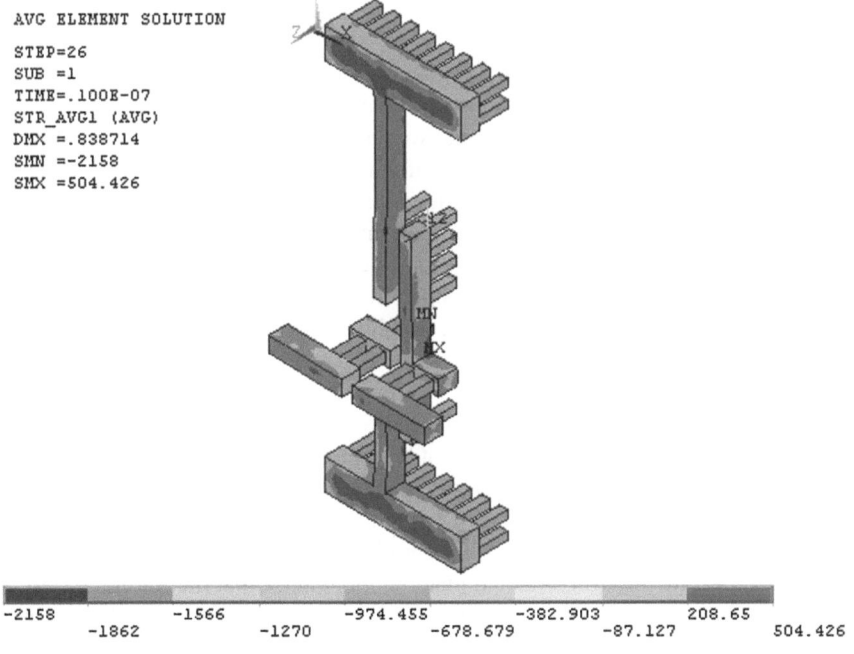

Fig. 2.31 Thermo-mechanical stress distribution in the Cu/SiLK structure (unit: MPa)

With varying line width on different parts of the interconnect structure, the relative barrier thickness and hence the structural confinement from the barrier and the dielectric are different [28]. In the Cu/SiLK structure, due to the large difference in CTE between Cu and SiLK, and that the CTE of SiLK is larger than Cu, the confinement effect provided by the Ta barrier at the regions with smaller inter-connect dimension (e.g., the contact or via regions) is no longer enough to produce the positive tensile stress. As a result, a negative compressive stress is produced, as can be seen from the blue and green regions in Fig. 2.31. This results in a com-bination of the positive tensile and negative compressive stresses on different parts of the interconnects and thus increases the stress gradient of the Cu/SiLK structure. The stress state of this interconnect structure becomes dominated by shear stresses and the reliability concern changes to interfacial delamination [46].

In summary, this section presents the increase in the total AFD with decreasing barrier layer thickness and the use of the low-κ materials, using the inverter circuit model as an example. The increase in the total AFD with decreasing barrier layer thickness is due to the increase in the thermo-mechanical stress gradient with a thinner barrier. The low-κ dielectric with a lower elastic modulus and a higher CTE as compared with SiO_2 reduces the structural confinement and renders a higher thermo-mechanical stress gradient, and thus a higher total AFD.

2.6 Summary

This chapter demonstrated the step by step construction of a complete 3D circuit model from its 2D IC layouts. For ease of demonstration, a simple inverter circuit with only the intra-block interconnects was used as an example. The method of conducting the transient electro-thermo-structural simulations using both Cadence (a circuit simulator) and ANSYS (a finite element tool) was also introduced.

The current density, temperature and thermo-mechanical stress, as well as the atomic flux divergences (AFDs) at any load step and any node of the model could be determined and plotted from the simulation results, and hence the EM weak spots of the interconnects of the circuit, which is the location with the highest total AFD could be identified.

Different from other 3D EM models in literature, the 3D model presented here was able to perform the transient analysis at circuit layout level. The simulation results successfully showed the transient temperature and stress responses to the changes in electrical loads and circuit temperature respectively. Therefore the model was capable to evaluate the EM reliability of the interconnects of the circuit with different loadings and under different operation conditions.

To demonstrate the structural and material modeling capability of the 3D circuit model, the effects of barrier layer thickness and low-κ dielectric on the EM reli-ability of the circuit were studied using the inverter circuit model. It was found that, although the use of the sub-5 nm barrier layers in combination with a low-κ

dielectric was beneficial for reducing the RC delay, a higher maximum total AFD was observed in the structures with a thinner barrier or with a low-κ dielectric. A shorter EM lifetime was expected when the barrier thickness scaled below 7.5 nm and/or when the SiO_2 ILD was replaced by a low-κ dielectric.

References

1. Complete timing signoff in the nanometer era, http://w2.cadence.com/whitepapers/timing_signoff_wp.pdf
2. Baker RJ (2004) CMOS circuit design—layout and simulation, revised 2nd edn. IEEE Press Series on microelectronic Systems, Wiley, London
3. Tan CM, He F (2009) 3D circuit model for 3D IC reliability study. In: 10th International conference on thermal, mechanical and multi-physics simulation and experiments in microelectronics and microsystems (EuroSimE), 2009, p 7
4. Tan CM, Hou Y, Li W (2007) Revisit to the finite element modeling of electromigration for narrow interconnects. J Appl Phys 102(2):033705-1-7
5. He F, Tan CM (2010) Circuit level interconnect reliability study using 3D circuit model. Microelectron Reliab 50(3):376–390
6. He F, Tan CM (2010) Modeling the effect of barrier thickness and low-κ dielectric on circuit EM reliability using 3D model. Microelectron Reliab 50(9–11):1327–1331
7. Alvino WM (1994) Plastics for electronics, materials, properties, and design applications. McGraw-Hill, New York, p 326
8. Chartered Semiconductor Manufacturing. Document 0.18um DROIT_06 June 07
9. Chung DDL (1995) Materials for electronic packaging. Butterworth-Heinemann, London, p 57
10. Pecht M (1999) Electronic packaging materials and their properties. CRC Press, Boca Raton, p 151
11. Dalleau D, Weide-Zaage K, Danto Y (2003) Simulation of time depending void formation in copper, aluminum and tungsten plugged via structures. Microelectron Reliab 43(9–11):1821–1826
12. Chen CT, Hsu T-S, Jeng R-J, Yeh H-C (2000) Enhancing the glass-transition temperature of polyimide copolymers containing 2,2-bipyridine units by the coordination of nickel malenonitriledithiolate. J Polym Sci Part A: Polym Chem 38(3):498–503
13. Dalleau D, Weide-Zaage K (2001) Three-dimensional voids simulation in chip metallization structures: a contribution to reliability evaluation. Microelectron Reliab 41(9–10):1625–1630
14. Ciptokusumo J, Weide-Zaage K, Aubel O (2009) Investigation of stress distribution in via bottom of Cu-via structures with different via form by means of submodeling. Microelectron Reliab 49(9–11):1090–1095
15. Weide-Zaage K, Dalleau D, Danto Y, Fremont H (2007) Dynamic void formation in a DD-copper-structure with different metallization geometry. Microelectron Reliab 47(2–3):319–325
16. Gonzalez JL, Rubio A (1997) Shape effect on electromigration in VLSI interconnects. Microelectron Reliab 37(7):1073–1078
17. What temperature should my processor be running at, http://www.computerhope.com/issues/ch000687.htm
18. Chen D, Li E, Rosenbaum E, Kang S-M (2000) Interconnect thermal modeling for accurate simulation of circuit timing and reliability. IEEE Trans Comp-Aided Des Integr Circuits Syst 19:197–250

19. Lasance C, Moffat C (2005) Advances in high-performance cooling for electronics. Electron Cooling 11(4), http://www.electronics-cooling.com/2005/11/advances-in-high-performance-cooling-for-electronics/
20. Cheng Y-K, Tsai C-H, Teng C-C, Kang S-M (2002) Electrothermal analysis of VLSI systems. Kluwer Academic Publishers, Dordrecht
21. Rzepka S, Banerjee K, Meusel E, Hu CM (1998) Characterization of self-heating in advanced VLSI interconnect lines based on thermal finite element simulation. IEEE Trans Compon Packag Manuf Tech Part A 21(3):406–411
22. Lloyd JR (1999) Electromigration in integrated circuit conductors. J Appl Phys 32(17):R109–R118
23. Schoen JM (1980) A model of electromigration failure under pulsed condition. J Appl Phys 51:508
24. Tao J, Cheung NW, Hu C (1994) Electromigration failure model for interconnects under pulsed and bidirectional current stressing. IEEE Trans Electron Devices 41(4):539–545
25. Adler V, Friedman EG (1996) Delay and power expressions for a CMOS inverter driving a resistive-capacitive load. In: IEEE international symposium on circuits and systems, vol 4, pp 101–104
26. Chen WY (2004) Home networking basics. Prentice Hall, Englewood Cliffs
27. Shen Y-L, Ramamurty U (2003) Temperature-dependent inelastic response of passivated copper films: experiments, analyses, and implications. J Vac Sci Tech B: Microelectron Nanometer Struct 21(4):1258–1264
28. Rhee S-H, Du Y, Ho PS (2003) Thermal stress characteristics of Cu/oxide and Cu/low-κ submicron interconnect structure. J Appl Phys 93(7):3926–3933
29. Love AE (1927) A treatise on the mathematical theory of elasticity. Cambridge University Press, Cambridge
30. Released 11.0 Documentation for ANSYS, Chapter 9. Submodeling
31. Wang E, Nelson T, Rauch R (2004) Back to elements—tetrahedra vs. hexahedra. In: International ANSYS Conference, 2004
32. Wang TH (2005) Submodeling analysis for path-dependent thermomechanical problems. Trans ASME J Electron Packag 127(2):135–140
33. Christou A (1993) Electromigration and electronic device degradation. John Wiley, New York
34. Roy A, Tan CM (2006) Experimental investigation on the impact of stress free temperature on the electromigration performance of copper dual damascene submicron interconnect. Microelectron Reliab 46(9–11):1652–1656
35. Li W, Tan CM, Hou Y (2007) Dynamic simulation of electromigration in polycrystalline interconnect thin film using combined Monte Carlo algorithm and finite element modeling. J Appl Phys 101(10):104314
36. Alam SM, Gan CL, Thompson CV, Troxel DE (2007) Reliability computer-aided design tool for full-chip electromigration analysis and comparison with different interconnect metallizations. Microelectron J 38(4–5): 463–473
37. Petrov N, Valverde C (2005) Process control and material properties of thin electroless Co-based capping layers for copper interconnects. In: Proceeding of SPIE, 2005, vol 6002, pp 01–011
38. Sukharev V (2005) Physically based simulation of electromigration-induced degradation mechanisms in dual-inlaid copper interconnects. IEEE Trans Comp-Aided Des Integr Circuits Syst 24(9):1326–1335
39. Tan CM, Roy A (2006) Investigation of the effect of temperature and stress gradients on accelerated EM test for Cu narrow interconnects. Thin Solid Films 504(1–2):288–293
40. Attardo MJ, Rosenberg R (1970) Electromigration damage in aluminum film conductors. J Appl Phys 41:2381–2386
41. Wang MT, Lin YC, Chen MC (1998) Barrier properties of very thin Ta and TaN layers against copper diffusion. Electrochem Soc 145:2538–2545

42. Ou KL, Wu WF, Chiou SY (2007) Enhancing the reliability of n+ −p junction diodes using plasma treated tantalum barrier film. Microelectron Eng 84:151–160
43. Lu X, Pyun JW, Li B, Henis N, Neuman K, Pfeifer K, Ho PS (2005) Barrier layer effects on electromigration reliability of Cu/low k interconnects. In: IEEE Proceedings of inter interconnect technology conference 2005, pp 33–35
44. Paik JM, Park H, Joo YC (2004) Effect of low-κ dielectric on stress and stress-induced damage in Cu interconnects. Microelectron Eng 71:348–357
45. Cheng YL, Wang YL, Chen HC, Lin JH (2006) Effect of inter-level dielectrics on electromigration in damascene copper interconnect. Thin Solid Films 494:315–319
46. Zhao J-H, Qi W-H, Ho PS (2002) Thermomechanical property of diffusion barrier layer and its effect on the stress characteristics of copper submicron interconnect structures. Microelectron Reliab 42:27–34
47. Tan CM (2010) Electromigration in ULSI interconnects. World Scientific, Singapore
48. Lee HJ, Christopher SL, Liu DW, Bauer BJ, Lin EK, Wu WL, Grill A (2004) Structural characterization of porous low-κ thin films prepared by different techniques using x-ray porosimetry. J Appl Phys l95:2355–2359
49. Chen F, Gill J, Harmon D, Sullivan T, Li B, Strong A, Rathore H, Edelstein D, Yang CC, Cowley A, Clevenger L (2004) Measurements of effective thermal conductivity for advanced interconnect structures with various composite low-κ dielectrics. In: IEEE International Reliability Physics Symposium, 2004, pp 68–73
50. Webb E, Witt C, Andryuschenko T, Reid J (2004) Integration of thin electroless copper films in copper interconnect metallization. J Appl Electrochem 34:291–300

Chapter 3
Comparison of EM Performances in Circuit and Test Structures

3.1 Introduction

As discussed in Chap. 1, 3D EM simulation is necessary as the circuit structure is indeed 3D in their actual physical implementation. Most of the 3D EM models in literature use the line-via test structures which are only part of the real circuit structure. To further demonstrate the necessity for complete circuit modeling, the comparison of the EM performances in a circuit structure and a standard line-via test structure is performed in this chapter. Both the EM test condition and the circuit operation condition are considered.

Modifications to the interconnect structures are also carried out, and the effects of different interconnect structures such as the change in the via/contact position and its number, the inter-transistor distance, the metal structure and layers, and the transistor finger number on the EM reliability of the circuit considered are examined.

3.2 Model Construction and Simulation Setup

Chapter 2 focused on the methodology of construction and simulation of 3D finite element circuit model and a circuit with an obvious variation in the current and voltage waveforms is chosen, namely the inverter circuit, so as to demonstrate the change in temperature and stress with varying loads. The main purpose of this chapter is to compare the difference in the EM performance of the circuit and the test structures. As the EM modeling and experiments of the test structures are usually performed under DC condition, a DC dominating circuit is more suitable than the inverter circuit.

Class-AB amplifier is a common building block for an analog circuit. To reduce the complexity of the work, only the output stage of a class-AB amplifier is used as an example. The schematic of the output stage of the class-AB amplifier studied in this work is shown in Fig. 3.1.

C. M. Tan and F. He, *Electromigration Modeling at Circuit Layout Level*,
SpringerBriefs in Reliability, DOI: 10.1007/978-981-4451-21-5_3, © The Author(s) 2013

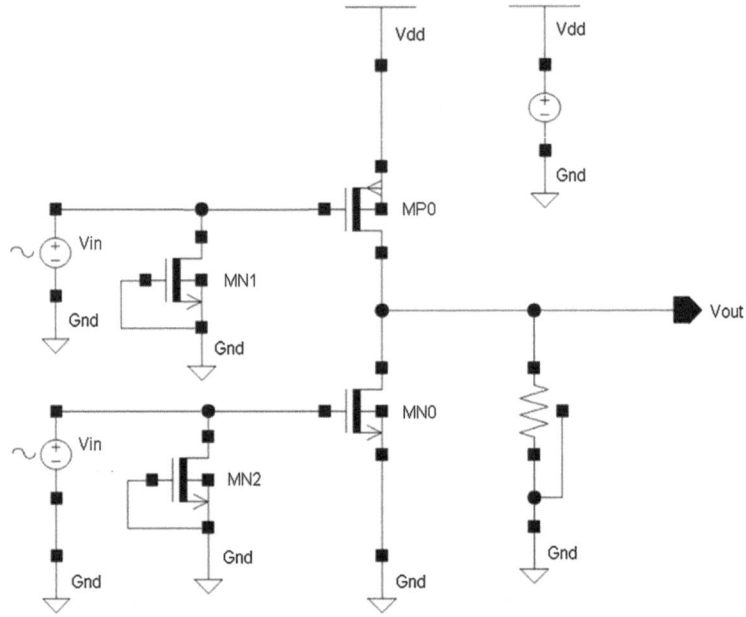

Fig. 3.1 The schematic of the output stage of a class-AB amplifier [1]

This output stage contains one PMOS (MP0) and one NMOS (MN0) as a functional block, and two short circuit protection transistors MN1 and MN2. As we are focusing on the functional block of the circuit, the short circuit protections MN1 and MN2 are not included in the 3D model as they do not affect the temperature and thermo-mechanical stress distributions of the model. To minimize the size of the 3D model, the sizes of MP0 and MN0 are kept small with a transistor width $W_t = 3$ μm. The supply voltage from V_{dd} is 3.3 V, and the input voltage V_{in} is 0.7 V. The circuit operates under small signal excitation and drives a resistive load of 1 kΩ [1]. The threshold voltage V_t of MP0 and MN0 is −0.461 and 0.483 V, respectively, as obtained from Cadence. The same frequency as in Chap. 2 is chosen as an example. Other loadings and operating frequencies can also be used as long as the circuit remains DC dominating. The source and drain currents and voltages of the transistors extracted from Cadence are used as the inputs to the 3D model in ANSYS.

Note here that only one PMOS and one NMOS are used for the model construction and simulation, in the remaining section of this chapter, "PMOS" refers to MP0 and "NMOS" refers to MN0.

Figure 3.2 shows the 3D model of the output stage of the class-AB amplifier and the line-via test structure [2]. As can be seen in Fig. 3.2b, the line-via test structure is part of the connection between the two transistors of the circuit structure. The rectangular region "cutout" from the circuit structure is the line-via

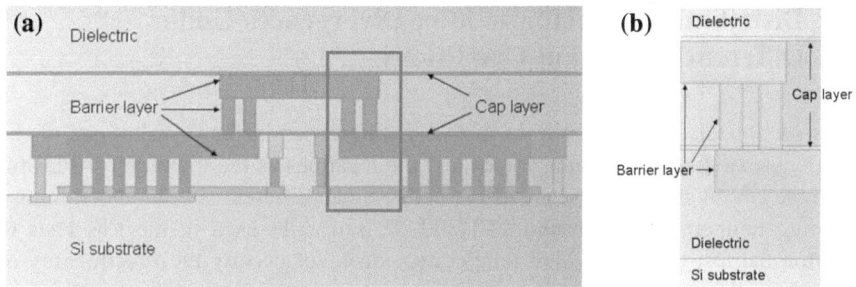

Fig. 3.2 *Side view* of **a** the 3D circuit structure and **b** the line-via test structure which is part of the circuit structure as shown in the *rectangular box* (Reprinted from, He and Tan [2], Copyright (2012), with permission from Elsevier)

test structure, and it is used to compare with the circuit structure. The structure is taken from the PMOS region as the PMOS is the main current flow path under this loading condition, and the amount of current flow in the PMOS is more than ten times larger than that in the NMOS.

In this modeling work, the following assumptions are made:

- The transistors are assumed to be reliable, and the focus of this work is on the EM reliability of the interconnects.
- Transistor diffusion region is treated as the heat source, and the input currents are applied at the contact/diffusion region interface of the 3D circuit structure
- Local current density at Metal 1 is extracted from the simulation results of the circuit structure, and it is used as the input in the simulation of the line-via test structure, so as to make sure that the two structures are operating under the same condition.
- Realistic circuit using 0.18 μm technology has six metal layers. To reduce the complexity of the model and for ease of comparison, only Metal 1, Metal 2, Via 12, and the contacts of the interconnect structure are considered, and the input voltage is applied at Metal 2.
- The skin depth at an operating frequency of 100 MHz is larger than the thicknesses of Metal 1, Metal 2, Via 12, and the contacts, and thus, the skin effect is ignored in this simulation.

Quadratic tetrahedral element SOLID 98 is used in the ANSYS simulation for both structures. The additional layers and the surrounding materials and their properties, the boundary conditions, and the setup for the transient electro-thermo-structural simulations are the same as the inverter circuit model in Chap. 2

3.3 Distributions of Atomic Flux Divergences Under
   Different Operation Conditions

This section discusses the similarities and differences in the distributions of the AFDs due to the three driving forces and the locations of the maximum AFDs for both the circuit structure and the test structure under different operation conditions.

High temperature of around 250–300 °C is usually used in the EM tests to shorten the test time. The high temperature does not destroy the functionality of the structure as the line-via test structure contains only the metallization and its surrounding materials with no circuit components involved. The simulation is first conducted at an EM test temperature of 300 °C.

At the end of one load cycle (i.e., 10 ns), the two structures have similar total AFD distributions as can be seen in Fig. 3.3. The locations of the maximum total AFD are found at the Metal 1/Via 12 interface (Site A) for both structures, and Region 1 has high total AFD values (i.e., above 8.05×10^9 atoms/μm^3 s for the circuit structure and 1.27×10^{10} atoms/μm^3 s for the line-via test structure). This result is consistent with the experimental results in literature [3, 4].

However, test temperature of 300 °C is too high for the circuit under actual operation because most of the circuit components are not able to function well at above 125 °C and the chip packaging will change its properties significantly above its glass transition temperature which is 260 °C for the commonly used packaging materials in IC packaging [5]. The simulation is performed again at a circuit operation temperature of 90 °C, and the results are shown in Fig. 3.4.

From Fig. 3.4, we can see that the total AFD distributions at 90 °C are different from those at 300 °C for both structures. For the circuit structure, the region with high total AFD values (i.e., above 5.70×10^7 atoms/$\mu m^3 \cdot s$) is now at the contact region (Region 2) instead of Region 1. The location of the maximum total AFD changes to the contact/Metal 1 interface (Site B) which is the location of the second highest total AFD at 300 °C in Fig. 3.3a. For the line-via test structure, the location of the maximum total AFD changes to the corner of Metal 1 (Site C).

As we can see, the locations of the maximum total AFD of the circuit structure and the line-via test structure are the same at 300 °C in this case study, and hence, the EM failure site in the circuit structure, if possible, will be the same as that obtained in the line-via test structure under the EM test condition. Therefore, no suspicion arises, and the extrapolation of the EM test data from 300 °C will then be conveniently performed. However, as can be seen in Fig. 3.4, the locations of the maximum total AFD of both structures are different between 300 and 90 °C, and as will be seen later in Table 3.1, the dominant driving forces for EM at 300 °C for both structures are different from that at 90 °C. Therefore, the EM weak spot and the EM lifetime obtained through the conventional extrapolation method will be incorrect. In fact, with different dominant failure physics, the conventional extrapolation method is not valid [6]. Further experimental verification will be interesting.

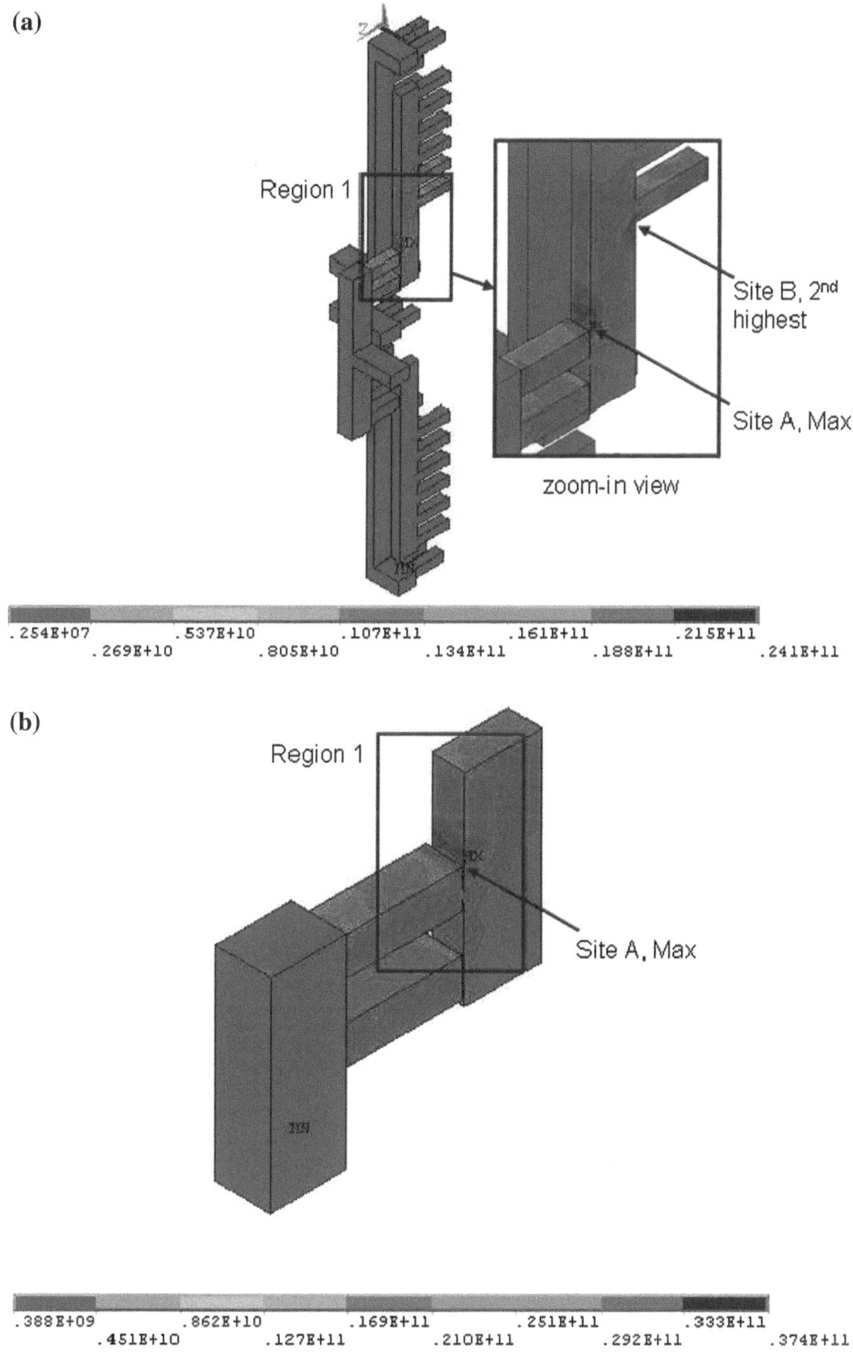

(a)

Region 1

Site B, 2nd highest

Site A, Max

zoom-in view

.254E+07 .537E+10 .107E+11 .161E+11 .215E+11
.269E+10 .805E+10 .134E+11 .188E+11 .241E+11

(b)

Region 1

Site A, Max

.388E+09 .862E+10 .169E+11 .251E+11 .333E+11
.451E+10 .127E+11 .210E+11 .292E+11 .374E+11

Fig. 3.3 Total AFD distributions of **a** the circuit structure and **b** the line-via test structure at 300 °C (unit: atoms/μm³ · s). The maximum AFD locations are found at the Metal 1/Via 12 interface for both structures (Reprinted from He and Tan [2] Copyright (2012), with permission from Elsevier)

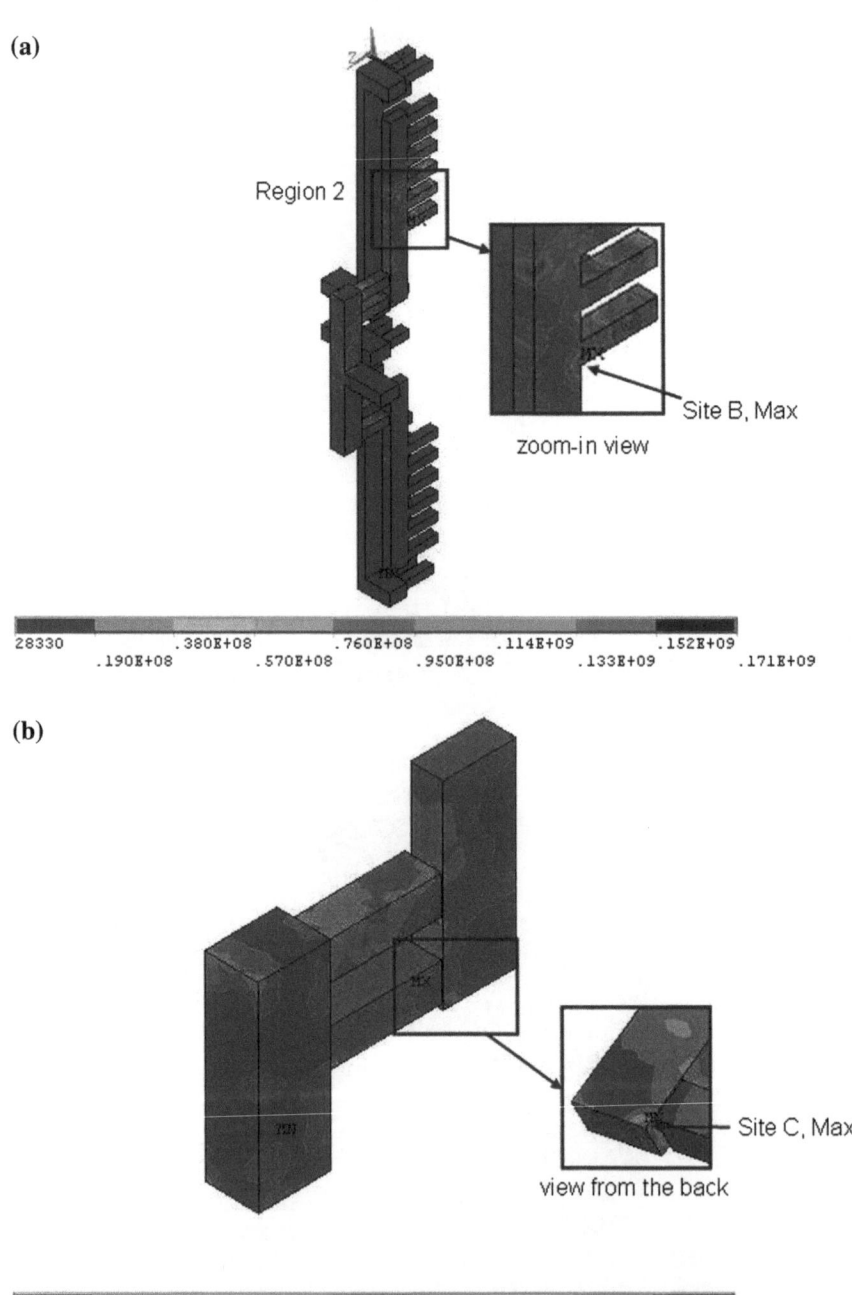

Fig. 3.4 Total AFD distributions of **a** the circuit structure and **b** the line-via test structure at 90 °C (unit: atoms/μm³ · s). The maximum AFD locations are different for the two structures (Reprinted from He and Tan [2], Copyright (2012), with permission from Elsevier)

Table 3.1 Percentage contribution of the three driving forces to the total AFD for the two structures under different operation conditions (Reprinted from He and Tan [2], Copyright (2012), with permission from Elsevier)

Test temperature (°C)		Percentage contribution		
		AFD_EWF	AFD_S	AFD_T
Circuit structure	300	1.95	33.32	64.73
	90	0.62	93.57	5.81
Line-via test structure	300	1.67	37.10	61.23
	90	0.51	97.64	1.85

The change in the distribution and the location of the maximum total AFD is due to the increase in the thermo-mechanical stress and stress gradient and, hence, the change in the dominant driving force for the total AFD at the circuit operation temperature. The AFD distributions due to electron wind force, thermo-mechanical stress gradient induced driving force, and temperature gradient induced driving force at the two test temperatures for both structures are shown in Figs. 3.5, 3.6, 3.7, and 3.8, respectively, and their percentage contribution to the total AFD is shown in Table 3.1.

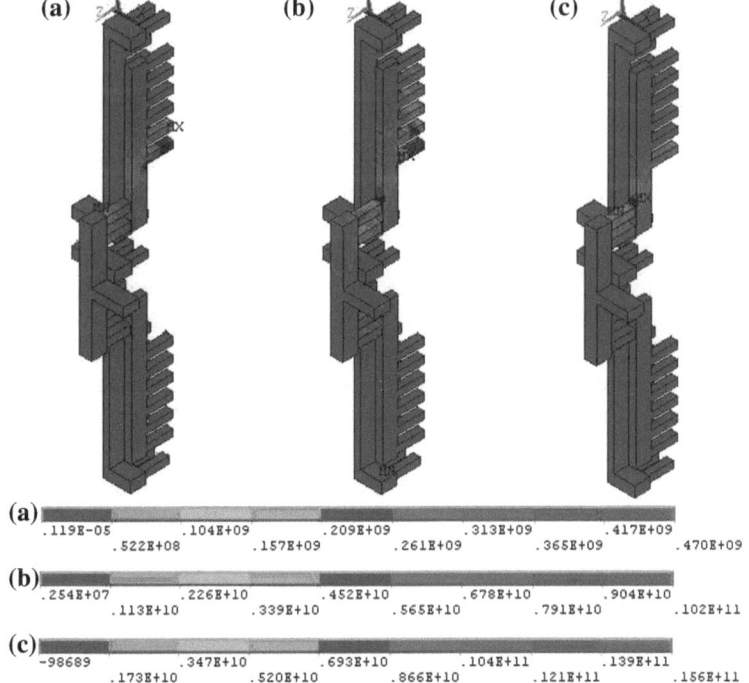

Fig. 3.5 AFD distributions due to **a** electron wind force, **b** thermo-mechanical stress gradient induced driving force, and **c** temperature gradient induced driving force of the circuit structure at 300 °C (unit: atoms/$\mu m^3 \cdot$ s) (Reprinted from He and Tan [2], Copyright (2012), with permission from Elsevier)

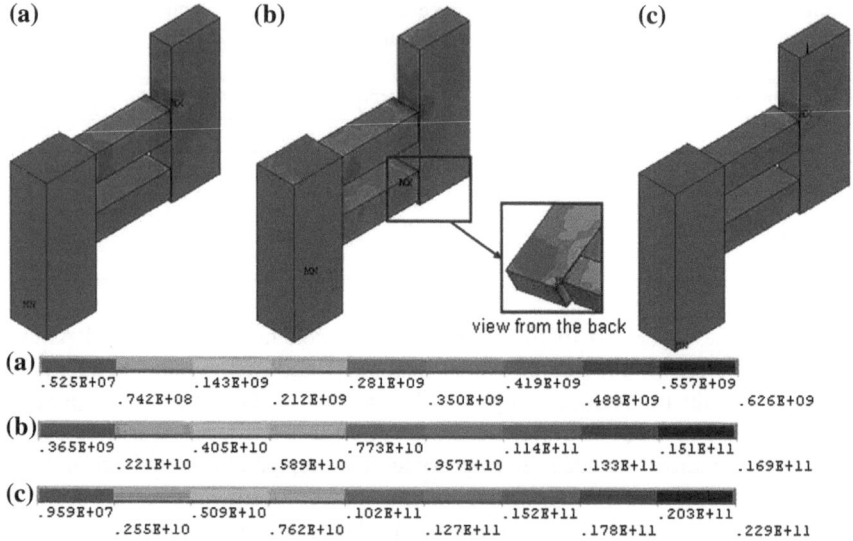

Fig. 3.6 AFD distributions due to **a** electron wind force, **b** thermo-mechanical stress gradient induced driving force, and **c** temperature gradient induced driving force of the line-via test structure at 300 °C (unit: atoms/$\mu m^3 \cdot$ s) (Reprinted from He and Tan [2], Copyright (2012), with permission from Elsevier)

The thermo-mechanical stress in the interconnects is directly proportional to the temperature difference between the interconnect temperature and the SFT of the metallization [7]. At the EM test temperature of 300 °C, the interconnect temperature is close to the SFT of 350 °C. The thermo-mechanical stress and stress gradient in the interconnects is relatively small (i.e., less than 100 MPa and 850 MPa/μm, respectively), and this results in a smaller percentage contribution of AFD_S to the total AFD when compared with AFD_T, as shown in Table 3.1. As a result, the total AFD distribution mainly follows the distribution of AFD_T. From Eqs. (2.4) and (2.7), AFD_T depends on the current density and temperature gradient of the interconnects. At 300 °C, term 2 in (2.7) is around 5 orders of magnitude larger than term 1; hence, the current density in the interconnects is the determining factor for AFD_T.

As shown in Fig. 3.9, current crowding occurs at Sites A and B of the circuit structure and high current density of more than 2.06×10^{10} pA/μm^2 is observed at these locations. The locations of the maximum AFD_T and hence the maximum total AFD are found at the place with the highest current density, that is, the Metal 1/Via 12 interface (Site A).

From Table 3.1, we can also see that AFD_S has significant contribution to the total AFD. The location of the second highest total AFD is at the place with the highest AFD_S, that is, the contact/Metal 1 interface (Site B). From Eqs. (2.5) and (2.8), AFD_S depends on the current density, thermo-mechanical stress gradient, and temperature gradient of the interconnects. At 300 °C, terms 1 and 3 in Eq.

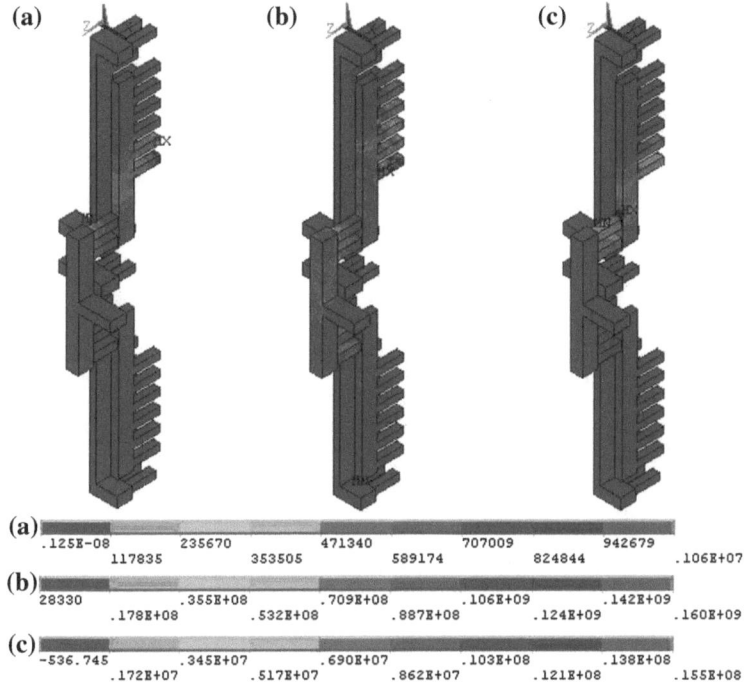

Fig. 3.7 AFD distributions due to **a** electron wind force, **b** thermo-mechanical stress gradient induced driving force, and **c** temperature gradient induced driving force of the circuit structure at 90 °C (unit: atoms/$\mu m^3 \cdot$ s) (Reprinted from He and Tan [2], Copyright (2012), with permission from Elsevier)

(2.8) are both around 6 orders of magnitude larger than term 2, and thus, AFD_S is determined by the current density and the product of the thermo-mechanical stress gradient and the temperature gradient.

The distributions of the temperature, temperature gradient and thermo-mechanical stress gradient of the circuit structure are shown in Fig. 3.10. At the end of one load cycle, Region 1 has a high temperature (i.e., above 300.23 °C) due to high current density (i.e., above 1.18×10^{10} pA/μm^2) at this region. The temperature gradient is higher at the bottom of the PMOS contacts (i.e., above 0.10 °C/μm), which is closer to both the heat source (the diffusion region) and the heat sink (the base of the metal plate). The temperature and current density of the PMOS contacts gradually decrease as we move away from Region 1, and the temperature gradient follows this trend. Site A has a small temperature gradient of less than 0.015 °C/μm due to the relatively uniform temperature distribution around Site A, as can be observed in Fig. 3.10a (i.e., the red color region with a temperature difference of less than 0.026 °C). The thermo-mechanical stress of the interconnect is dominated by the intrinsic stress of the metallization as explained in Sect. 2.4.4, but its value is more than five times smaller than that in Fig. 2.24 when the difference between the interconnect temperature and the SFT drops from

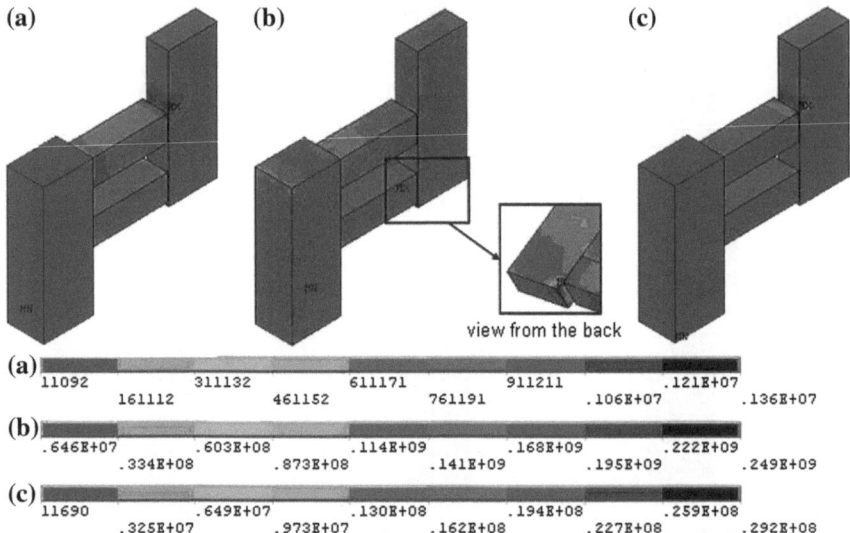

Fig. 3.8 AFD distributions due to **a** electron wind force, **b** thermo-mechanical stress gradient induced driving force, and **c** temperature gradient induced driving force of the line-via test structure at 90 °C (unit: atoms/$\mu m^3 \cdot$ s) (Reprinted from He and Tan [2], Copyright (2012), with permission from Elsevier)

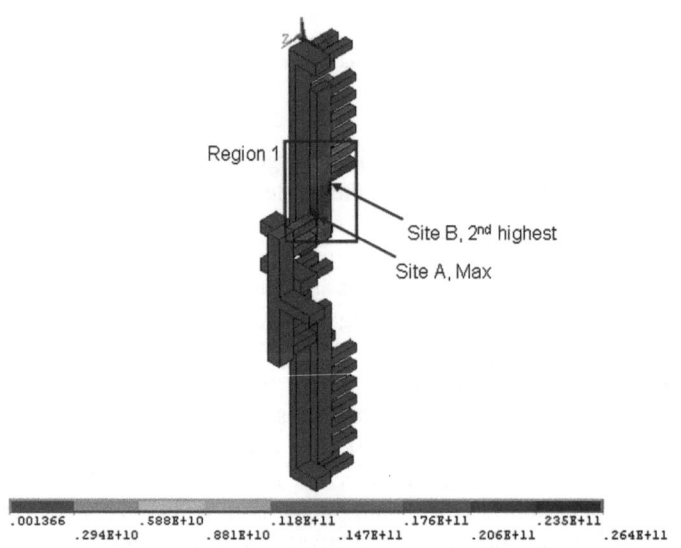

Fig. 3.9 Current density distribution of the circuit structure at 300 °C (unit: pA/μm^2). The arrow and the box indicate the maximum current density location and the high current density region, respectively (Reprinted from He and Tan [2], Copyright (2012), with permission from Elsevier)

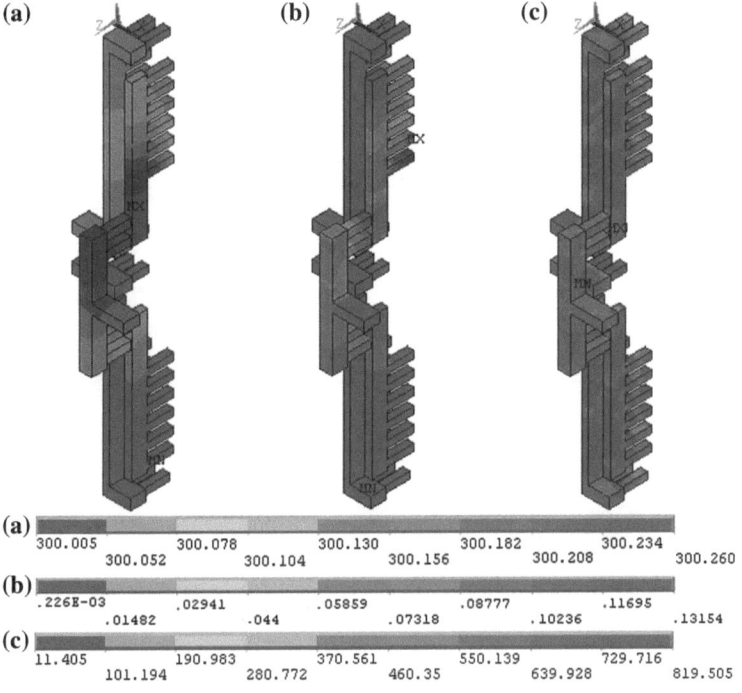

(a)

| 300.005 | | 300.078 | | 300.130 | | 300.182 | | 300.234 | |
| 300.052 | | 300.104 | | 300.156 | | 300.208 | | 300.260 |

(b)

| .226E-03 | | .02941 | | .05859 | | .08777 | | .11695 | |
| .01482 | | .044 | | .07318 | | .10236 | | .13154 |

(c)

| 11.405 | | 190.983 | | 370.561 | | 550.139 | | 729.716 | |
| 101.194 | | 280.772 | | 460.35 | | 639.928 | | 819.505 |

Fig. 3.10 Distributions of **a** temperature (unit: °C), **b** temperature gradient (unit: °C/μm), and **c** thermo-mechanical stress gradient (unit: MPa/μm) of the circuit structure at 300 °C

260 to 50 °C. This causes the model to have a smaller thermo-mechanical stress gradient of less than 850 MPa/μm at 300 °C when compared to the case at 90 °C, as shown in Fig. 3.10c.

With similar stress gradients at Sites A and B, though the current density at Site A is around 1.5 times higher than that at Site B, the temperature gradient at Site B is more than three times higher than that at Site A. The combined effect of terms 1 and 3 in Eq. (2.8) causes AFD_S at Site B to be slightly higher (i.e., around 1.11 times) than that at Site A.

The same analysis applies to the simple line-via test structure. Figure 3.11 shows the distributions of the current density, temperature gradient and thermo-mechanical stress gradient of the line-via test structure at 300 °C. For this structure, the Metal 1/Via 12 interface (Site A) also has the highest current density, and thus, the locations of the maximum AFD_T and the maximum total AFD are found at Site A. The absence of the contact region causes the maximum AFD_S to fall at the corner of Metal 1 (Site C) with the highest stress gradient and temperature gradient product. With high current density of more than 2.21×10^{10} pA/μm^2, Site A still has relatively high AFD_S of above 2.51×10^{10} atoms/μm^3 · s.

In addition, when the input current rises and reaches the current density level used in the EM test, the percentage contribution of AFD_S to the total AFD

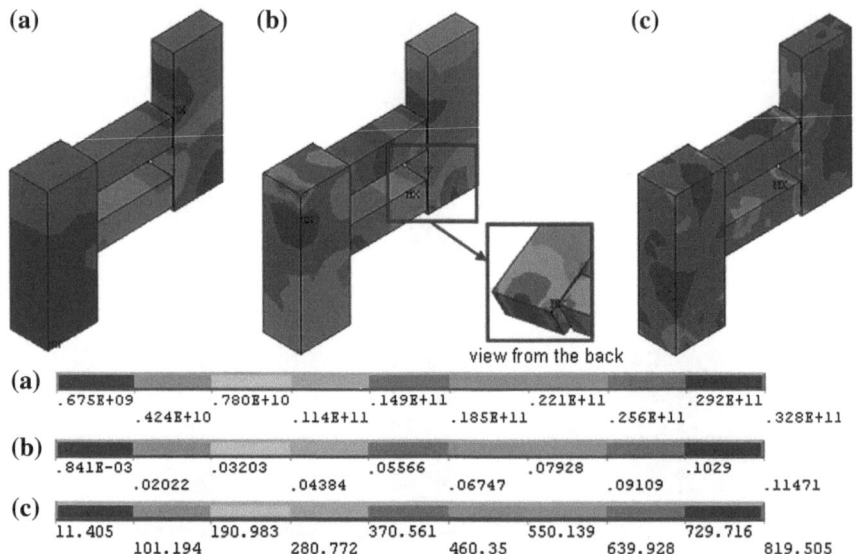

Fig. 3.11 Distributions of **a** current density (unit: pA/μm²), **b** temperature gradient (unit: °C/μm), and **c** thermo-mechanical stress gradient (unit: MPa/μm) of the line-via test structure at 300 °C

increases and overrides that of AFD_T. The percentage contribution of term 3 (i.e., the current density determining term) to AFD_S in Eq. (2.8) also increases. This causes a change of the location of the maximum AFD_S to Site A for both structures. Now, the location of the maximum AFD_T and the maximum AFD_S are both at Site A, and it is the location of the maximum total AFD.

At the circuit operation temperature of 90 °C, the distributions of the current density and temperature gradient for both structures are similar to that at 300 °C as the current and voltage loads under the two conditions are the same. The distributions of the thermo-mechanical stress and stress gradient for both structures are similar to that at 300 °C due to their similar intrinsic stress distributions.

However, the large difference between the interconnect temperature and the SFT causes around five times increases in the magnitudes of the thermo-mechanical stress and stress gradient in the interconnects at 90 °C when compared to that at 300 °C. With the same current loads, AFD_S becomes the dominant factor for the total AFD. Therefore, at 90 °C, the total AFD distribution and the maximum total AFD location depend on that of AFD_S. Term 1 in Eq. (2.8) is 8 and 1 order of magnitude larger than the other two terms, respectively. AFD_S at 90 °C mainly depends on the product of the thermo-mechanical stress gradient and the temperature gradient, but no longer on the current density.

As a result, the location of the maximum AFD_S of the circuit structure changes to Site B, and the region with high values of AFD_S (i.e., above 5.32×10^7 atoms/μm³ · s) moves to the contact region (Region 2) with high temperature gradient (i.e., above 0.10 °C/μm). For the line-via test structure, as the

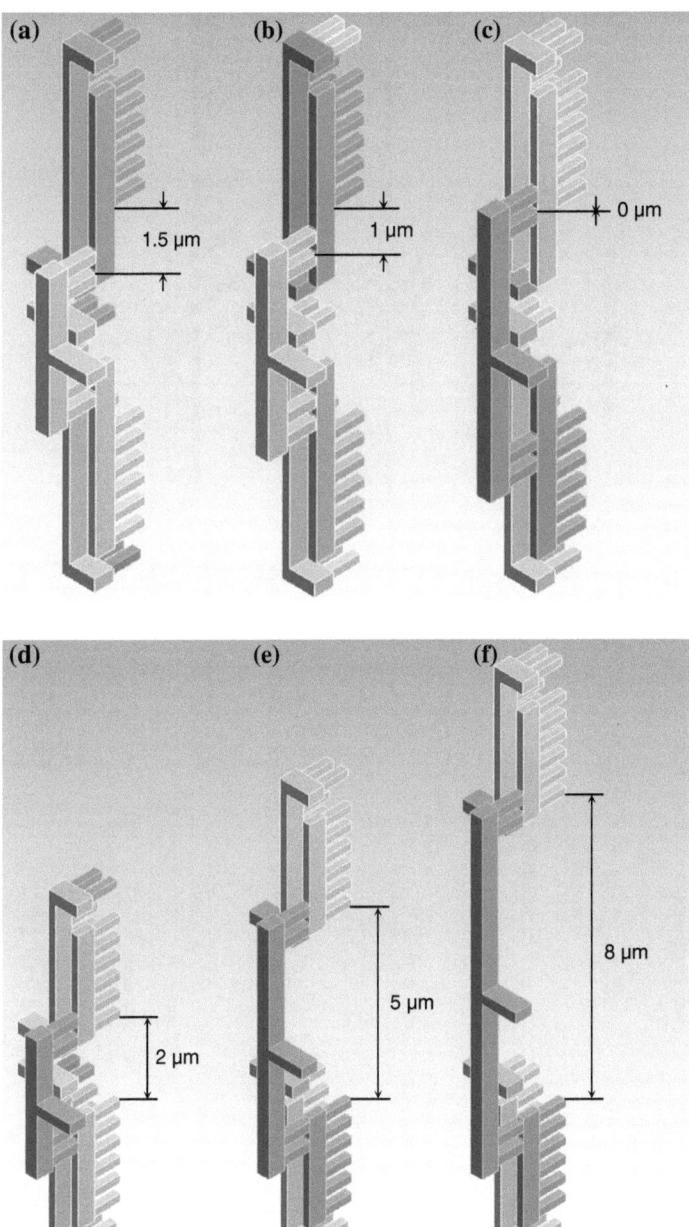

Fig. 3.12 The circuit models with different interconnect structures **a** the vias positioned far away from the contact region, **b** the vias positioned nearer to the contact region, **c** the vias positioned directly above the contact region, **d** an inter-transistor distance of 2 μm, **e** an inter-transistor distance of 5 μm, **f** an inter-transistor distance of 8 μm, **g** three contacts positioned directly below the vias, **h** three contacts positioned at the other end of Metal 1, **i** metal 1 used as the output line, **j** six metal layers positioned directly above the transistor, and **k** six metal layers positioned at a distance of 5 μm away from the transistor. Only the copper interconnects are shown for clarity

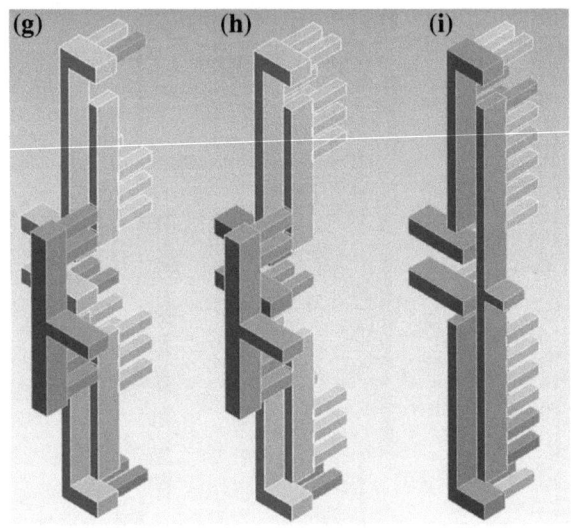

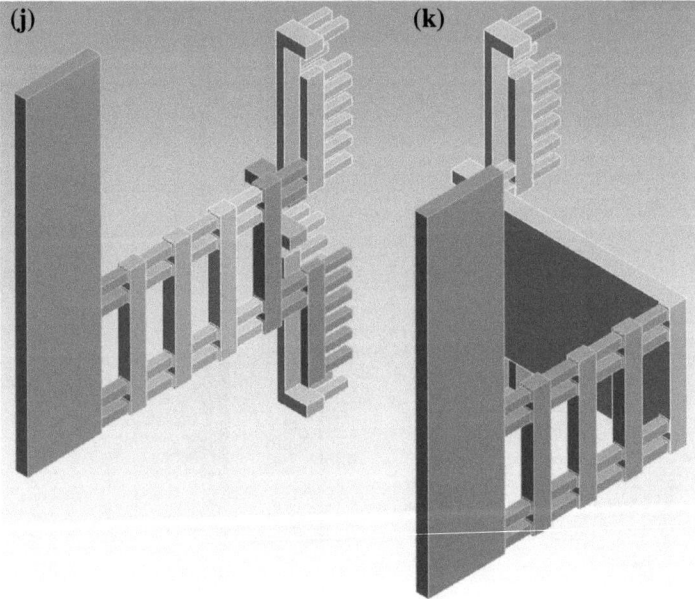

Fig. 3.12 (continued)

total AFD is now mainly determined by AFD_S, the location of the maximum total AFD is the same as that of AFD_S, and it is at the corner of Metal 1 (Site C). Site A no longer has the highest AFD_S even though it still has the highest current density.

In summary, while the circuit structure and the line-via test structure have the same maximum total AFD locations at the EM test temperature, they do not show the same location at the circuit operation temperature as the total AFD is no longer mainly determined by the current density. This can cause a misinterpretation of the void nucleation location when using the line-via test structure to model the EM reliability of the circuit.

Although the above results may be specific to this circuit, they do show a difference in the void nucleation location for the two structures under the different conditions. In other words, the EM tests based on the line-via test structure is only to validate the reliability of the interconnect technology with respect to EM, but not necessary the EM reliability of an IC.

3.4 Effects of Interconnect Structures on Circuit EM Reliability

Understand the different EM mechanisms between the circuit and test structures, we now study the effects of interconnect structures on the circuit EM reliability by modifying the circuit model.

Based on the model in Fig. 3.2a, 11 models with the following changes in the interconnect structures are built, as shown in Fig. 3.12a–k [8].

1. *Change in the position of the vias*: The vias are moved from the extreme end of Metal 1 (Fig. 3.12a) to the region closer to the contacts (Fig. 3.12b) and directly above the contacts (Fig. 3.12c).
2. *Change in the inter-transistor distance*: The inter-transistor distance (i.e., the distance between the diffusion regions of PMOS and NMOS) is varied from 2 μm (i.e., minimum distance, Fig. 3.12d) to 5 μm (Fig. 3.12e) and then to 8 μm (Fig. 3.12f).
3. *Change in the number and the position of the contacts*: The number of contacts at the source and drain regions is reduced from 6 in Fig. 3.12d to 3, with the contact position directly below the vias (Fig. 3.12g) and at the other end of Metal 1 (Fig. 3.12h).
4. *Change in the metal structure at the output line*: Metal 1 is used as the output line instead of Metal 2 (Fig. 3.12i).
5. *Change in the number and the position of the metal layers*: In realistic circuit, the output line of the transistor may connect to the next stage directly via Metal 2 or Metal 1 (e.g., to another transistor, like in Fig. 3.12a–i), or via Metal 1 to Metal 6 (e.g., to a capacitor or inductor, or to the output pad). The connection may be directly above the transistor (Fig. 3.12j), or at a distance away from the transistor (Fig. 3.12k). For easy comparison purpose, similar structures are used in Fig. 3.12j, k.

Table 3.2 Maximum total AFD for different interconnect structures

Structural changes	Maximum total AFD (×10^8 atoms/μm^3·s)			Remarks
Distance between the vias and the contacts	1.5 μm — 1.71	1 μm — 1.52	0 μm — 1.09	The closer the vias to the contacts, the smaller the max AFD
Inter-transistor distance	2 μm — 1.09	5 μm — 1.04	8 μm — 1.04	Insignificant effect
Number and position of the contacts	6 — 1.09	3 — 1.36	Near to via — 1.36, Far from via — 1.72	Fewer contacts increase the max AFD, especially when the distance between the contacts and the vias is larger
Metal structure at the output line	Metal 1 + 2 — 0.159	Metal 1 only — 1.09		1 + 2 structure shows a lower max AFD and an even total AFD distribution
Number and position of the metal layers	2 — 1.09	6 — 1.54	Above Metal 2 — 1.54, 5 μm away — 1.36	Higher number of metal layers increases the max AFD. The increase in max AFD is smaller when the additional metal layers are added further away from the transistor

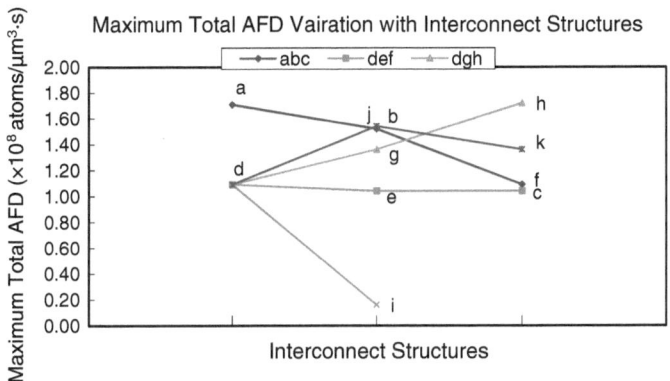

Fig. 3.13 Maximum total AFD for different interconnect structures

All the layouts are checked to comply with the design rules. The functionality of the circuit for the above structures remains unchanged as verified by Cadence. The simulation is carried out under the circuit operation of 90 °C with the same setup as in the previous section. The current loads are applied at the contact/diffusion region interface, and the voltage loads are applied at the top metal layer (Metal 2 for Fig. 3.12a–i and Metal 6 for Fig. 3.12j, k). The maximum total AFDs of the above structures at the end of one load cycle are shown in Fig. 3.13 and summarized in Table 3.2.

The change in the value and location of the maximum total AFD of the different interconnect structures may be explained as follows.

Under the circuit operation condition, AFD_S has more than 90 % contribution to the total AFD among the three driving forces. The total AFD distribution and the location of the maximum total AFD of the model are mainly determined by that of AFD_S. The location of the maximum AFD_S is found at the place where the product of the thermo-mechanical stress gradient and the temperature gradient is the highest, and it is also the location of the maximum total AFD.

The rise in temperature of the models at the end of one load cycle is less than 0.45 °C, and the thermo-mechanical stress of the interconnects is mainly determined by the intrinsic stress. The values of the maximum thermo-mechanical stress and stress gradient of the structures shown in Fig. 3.12a–k are similar due to their similar intrinsic stresses. The values and the locations of the maximum AFD_S and total AFD are therefore mainly affected by the temperature gradients of the models.

For more detailed discussion on the subject matter, please refer to our paper in [8].

1. The location of the maximum total AFD for Fig. 3.12a is at the contact/Metal 1 interface. The change in the position of the vias causes a change in the value and distribution of the current density. In Fig. 3.12a, there are two current crowding places, namely the Metal 1/Via 12 interface and the contact/Metal 1 interface. When the distance between the vias and the contacts decreases, the

crowding effect reduces. This causes a decrease in the current density with decreasing via/contact distance, especially when the vias are directly above the contacts. The reduction in the current density reduces the Joule heating and the temperature gradient of the model. Therefore, Fig. 3.12b and c have a smaller maximum AFD_S and hence a smaller maximum total AFD when compared to Fig. 3.12a (11.11 and 36.26 % reduction, respectively). A smaller maximum total AFD means a longer void nucleation time [9]. As the void nucleation time occupies, a large portion of the EM lifetime [10], increasing the void nucleation time can effectively enhance the EM lifetime of the circuit, and Fig. 3.12b and c are expected to have a longer EM lifetime than Fig. 3.12a.

Figure 3.12b has the same maximum total AFD location as Fig. 3.12a. For Fig. 3.12c, the contact/Metal 1 interface is no longer the current crowding region when the vias are directly above the contacts. The temperature gradient at this location decreases by more than five times. The locations of the maximum AFD_S and total AFD change to the bottom of the contact where the product of the stress gradient and the temperature gradient is the highest. Besides the effect due to current crowding, the temperature gradient at the Metal 1/Via 12 interface increases by more than two times as the vias are now directly above the heat source (the diffusion region) and the heat sink (the meal plate base). The locations of the second highest AFD_S and total AFD are found at the Metal 1/Via 12 interface.

2. The increase in the inter-transistor distance from 2 to 5 μm causes around 0.10 % reduction in the circuit temperature and temperature gradient as the heat sources are further apart. The reduction in the temperature and the temperature gradient gives rise to a 4.59 % reduction in the value of the maximum total AFD. The distributions of the temperature gradient and total AFD, and the locations of the maximum total AFD for Fig. 3.12d and e are similar to Fig. 3.12c due to their similar interconnect structures. This circuit has only one main heat generating transistor as the amount of current flow in NMOS is more than ten times smaller than in PMOS. Hence, the effect of inter-transistor distance is insignificant, and separating the two transistors further apart has very limited impact on the circuit temperature and temperature gradient. No further reduction in the value of the maximum total AFD is observed when the inter-transistor distance increases from 5 to 8 μm.

3. When the number of contacts is reduced from 6 to 3, the increase in the current density at each contact for Fig. 3.12g and h causes a rise in the temperature and temperature gradient of the model [11]; 24.77 and 57.80 % rise in the values of the maximum total AFD are observed for Fig. 3.12g and h (i.e., the three contacts structure), respectively, when compared with Fig. 3.12d (i.e., the six contacts structure), implying a shorter void nucleation time with fewer contacts. This result is expected and it agrees with the reported experimental observations in Lin et al. [3, 4]. Figure 3.12h has an even higher maximum total AFD than Fig. 3.12g due to the change in the position of the contacts (i.e., further away from the vias), and this corresponds to the observations in (1). The

location of the maximum AFD for Fig. 3.12h is different from Fig. 3.12a as the increase in the temperature gradient at the contact bottom has a larger impact on AFD_S than the increase in current density at the contact/Metal 1 interface. The locations of the maximum AFD_S and total AFD are found at the contact bottom.

4. In all the previous structures, the total AFD is always higher (i.e., above 1.00×10^8 atoms/μm^3) at the contacts nearer to the current crowding site (i.e., the vias at the output line) and lower (i.e., below 1.00×10^7 atoms/μm^3) at the contacts further away. When Metal 1 is used as the output line, the removal of the Metal 2/Via 12 structure eliminates the current crowing at this location. The distributions of the current density and temperature gradient of Fig. 3.12i at each contact of both the source and drain interconnects are uniform. The total AFD distribution is also uniform at each contact regardless of the contact position. The avoidance of the current crowding results in an 85.41 % drop in the value of the maximum total AFD for Fig. 3.12i when compared to that of Fig. 3.12d. The location of the maximum total AFD is found at the contact connecting to the source interconnects where the voltage is higher than at the drain interconnects (around 3.30 V and around 1.30 V, respectively).

5. Adding more metal/via layers causes an increase in the temperature gradient in the z-direction due to a larger distance from the metal top to the heat source, and thus, the maximum total AFD for this structure increases. The total AFD distribution and the location of the maximum total AFD for Fig. 3.12j are almost the same as Fig. 3.12d; therefore, the trends in (1)–(3) (where only two metal layers are considered) apply to the structures up to six metal layers.

When the metal/via stacks are added at a distance away from the transistor (i.e., away from the heat source), the impact of the additional metal/via layers on the temperature gradient of the circuit reduces. This results in a smaller increase in the maximum total AFD for Fig. 3.12k when compared with Fig. 3.12j (24.77 and 41.28 %, respectively). The maximum total AFD further reduces to 1.30×10^8 atoms/$\mu m^3 \cdot s$ when the width of the additional metal layers is doubled, and this is due to the decrease in current density with wider metal lines and more vias.

In summary, this section presents the impacts of interconnect structures on the EM reliability of the circuit. From the simulation results, it is observed that we can improve the EM lifetime of the circuit by placing the vias directly above the contact region, using more contacts and more uniform interconnect structures, or adding the Metals 3 to 6 connection away from the transistors. The observations are consistent with the experimental results in literature.

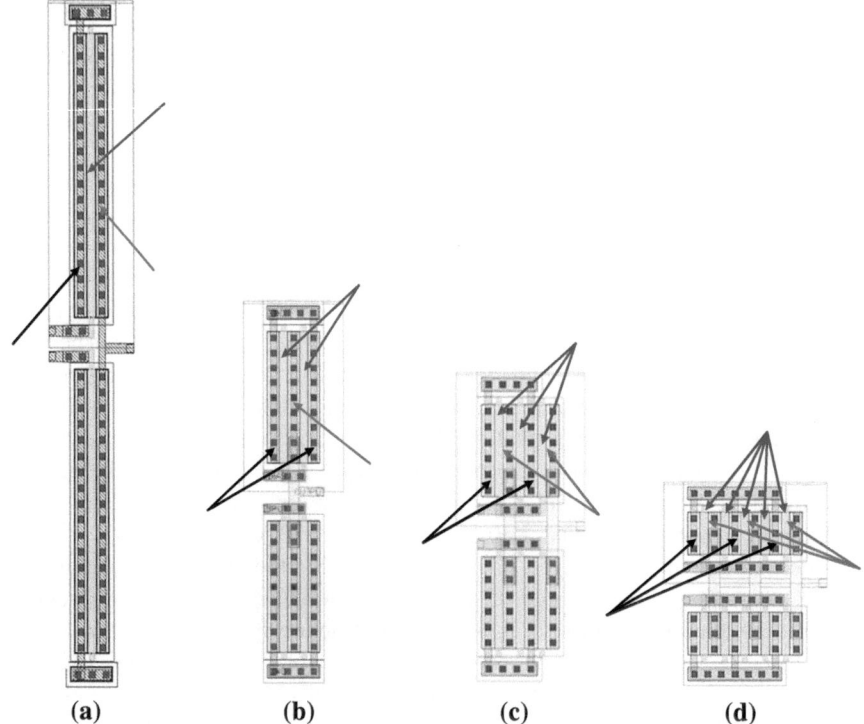

Fig. 3.14 Circuit layouts at the output stage of a class-AB amplifier with different finger numbers. The red squares are the contacts. The *black* and *red arrows* indicate the source and drain interconnects, respectively, and the *blue arrows* indicate the fingers (for PMOS only) **a** finger number = 1, contact number on source interconnects = 18, drain interconnect = 18, **b** finger number = 2, contact number on source interconnects = 18, drain interconnect = 9, **c** finger number = 3, contact number on source interconnects = 12, drain interconnect = 12, and **d** finger number = 5, contact number on source interconnects = 9, drain interconnect = 9 (Reprinted from He and Tan [12], Copyright (2012), with permission from Elsevier)

3.5 Effects of Transistor Finger Number on Circuit EM Reliability

The effects of transistor finger number on the EM reliability of the circuit are studied in this section based on the circuit model in Fig. 3.2a. In order to have at least two contacts on each finger, the transistor with a larger W_t of 9 μm is used (i.e., more contacts than the model in Fig. 3.2a). Figure 3.14 shows four circuit layouts with different number of fingers [12]. Based on the layout shown in Fig. 3.14a, the models with the number of contacts on the source and drain interconnects vary from 1 to 18 are constructed. The simulations are performed on these models.

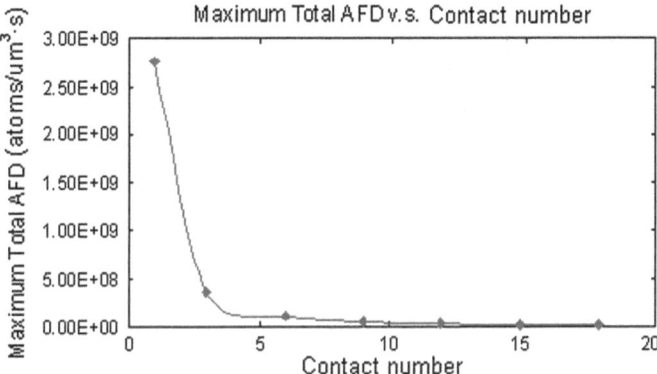

Fig. 3.15 Change in the maximum total AFD with the number of contacts on the source and drain interconnects of Layout 3.14a (Reprinted from He and Tan [12], Copyright (2012), with permission from Elsevier)

Table 3.3 The number of contacts on the source and drain interconnects for Layouts 3.14a–d (Reprinted from He and Tan [12], Copyright (2012), with permission from Elsevier)

Layout	Finger number	Contact number	
		Source	Drain
3.14(a)	1	18	18
3.14(b)	2	18	9
3.14(c)	3	12	12
3.14(d)	5	9	9

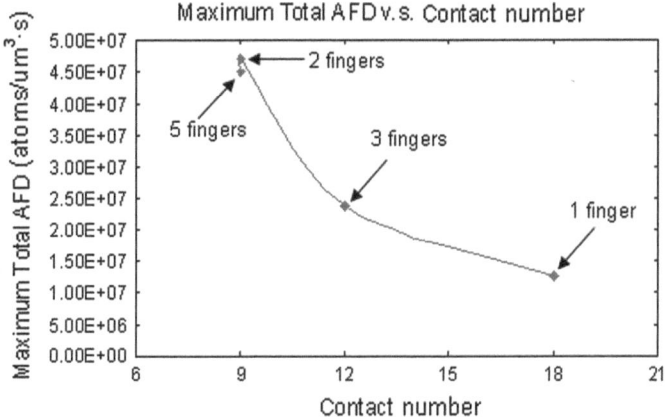

Fig. 3.16 Relationship between the finger number, the maximum total AFD, and the smaller number of contacts on the source and drain interconnects (Reprinted from He and Tan [12], Copyright (2012), with permission from Elsevier)

As illustrated in Fig. 3.15, an increase in the maximum total AFD is observed with decreasing number of contacts on the source and drain interconnects for the layout shown in Fig. 3.14a, and the effect of the number of contacts becomes significant when the contact number goes below 4.

The locations of the maximum total AFD of the four layouts with different number of fingers are found to be similar. Due to the limitation in the metal line length, the transistor with more fingers tends to have fewer contacts on the source and drain interconnects. As shown in Table 3.3, there are 12 contacts on the source or drain interconnects of Layout 3.14c with 3 fingers when compared to 18 contacts for Layout 3.14a with 1 finger, and the number further decreases to 9 for Layout 3.14d with 5 fingers. Due to the decrease in the number of contacts, an increase in current density and the maximum total AFD is observed with increasing number of fingers, as illustrated in Fig. 3.16. The increase in the maximum total AFD is not very significant as the smaller number of contacts on the source and drain interconnects for the four models is all larger than 4 as can be seen in Table 3.3.

Layout 3.14b with two fingers has two metal lines for the source and one metal line for the drain of a transistor. There are 18 contacts on the source interconnects and only nine contacts on the drain interconnect. As such, the current density at the drain region becomes almost double than that at the source region, and this increases the maximum total AFD of Layout 3.14b. Layout 3.14b therefore has a higher maximum total AFD when compared to Layout 3.14a even though they have the same number of contacts on the source interconnect, and this value is even higher than Layout 3.14c with 12 contacts on the source or drain interconnect, as can be seen in Fig. 3.16 and Table 3.3. In fact, Layout 3.14b shows almost

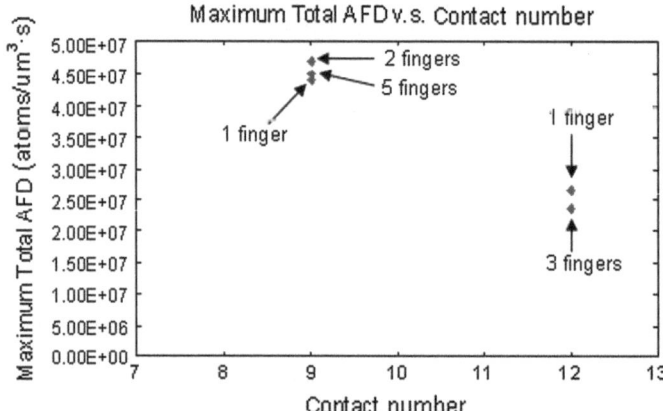

Fig. 3.17 Relationship between the finger number and the maximum total AFD for a given number of contacts on the source and drain interconnects. The impact of finger number becomes insignificant when the number of contacts on the source and drain interconnects is the same (Reprinted from He and Tan [12], Copyright (2012), with permission from Elsevier)

Fig. 3.18 Layout of a waffle
transistor [13]

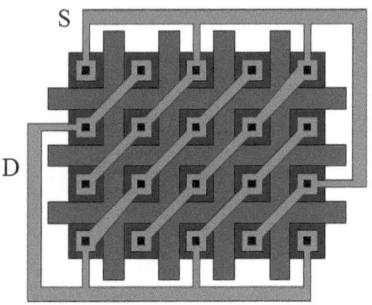

the same maximum total AFD as Layout 3.14d, as both of them have nine contacts on their drain interconnects.

From Fig. 3.16, we can see that the maximum total AFD depends on the smaller number of contacts on either the source or the drain interconnect instead of the number of fingers. For a given number of contacts on the source and drain interconnects, the impact on the maximum total AFD due to number of fingers is insignificant as can be seen in Fig. 3.17.

In summary, the increase in the number of fingers can cause a decrease in the total number of contacts due to the limitation in the metal line length of the source and drain interconnects for the case of small transistor, and this is not favorable as it increases the average current density and the total AFD. On the other hand, the difference in the total number of contacts with different finger number is insignificant for large transistors due to their longer metal line lengths. The maximum total AFD is almost independent on finger number. However, looking at the finger number and the contact number on the source and drain interconnects in Table 3.3, in order to obtain an even distribution in current density and the total AFD, it is better to use odd number of fingers so as to achieve the same number of contacts on both the source and the drain interconnects.

Another option in the case of large transistor is the use of waffle layout. An example of the waffle layout is shown in Fig. 3.18.

Comparing with the structure in Fig. 3.12, the uneven distribution of the contacts on the source and drain interconnects in the waffle layout (e.g., varying from 1 to 4 for the example in Fig. 3.18) creates significant current crowding and high temperature gradient on the interconnects with only one contact, as indicated by the circle in Fig. 3.19. The contact at this location tends to fail much faster than the others because this contact has a much larger maximum total AFD as can be seen in Fig. 3.20. Hence, the waffle layout is not favorable from the reliability point of view.

(a)

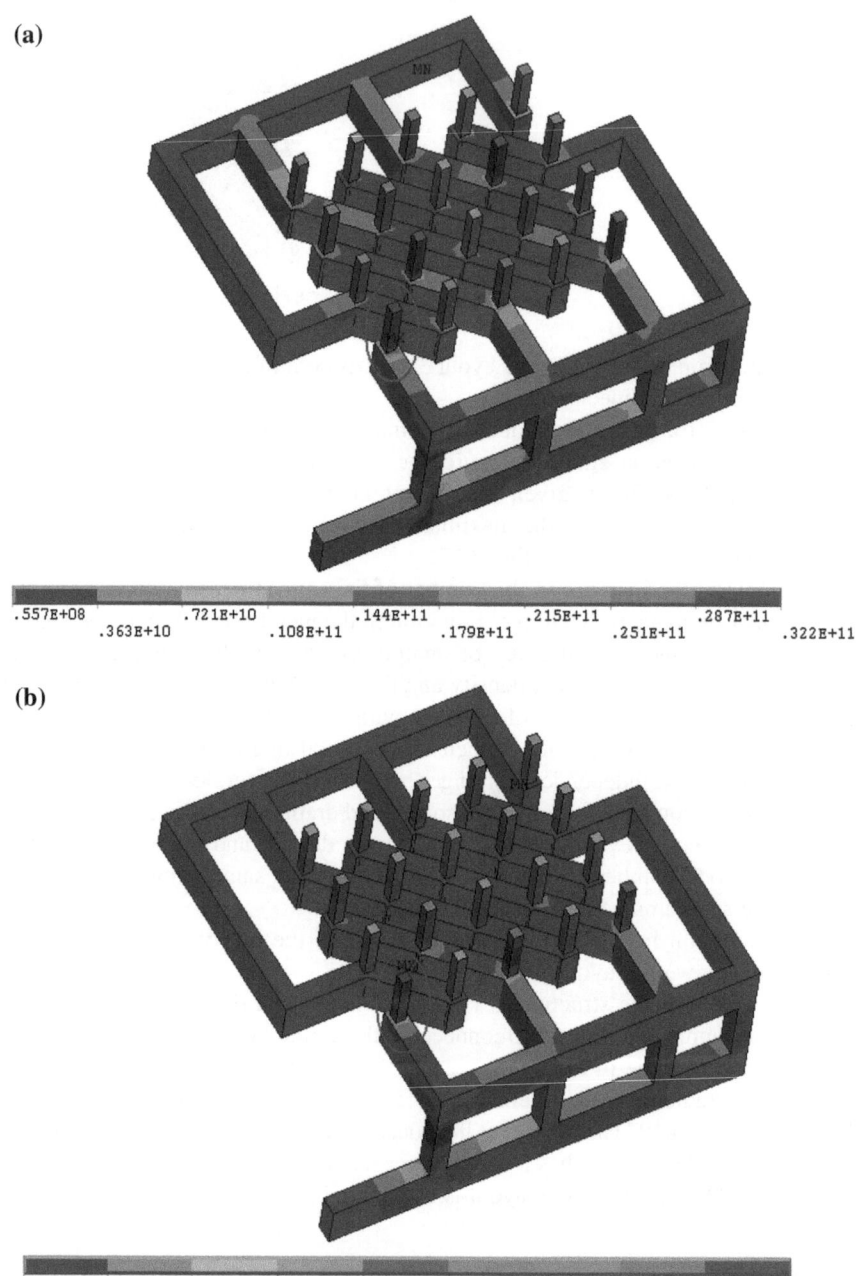

```
.557E+08              .721E+10              .144E+11              .215E+11              .287E+11
          .363E+10              .108E+11              .179E+11              .251E+11              .322E+11
```

(b)

```
.178E-03              .1107                 .22122                .33174                .44226
          .05544                .16596                .27648                .3870                 .49752
```

Fig. 3.19 Distributions of **a** current density (unit: pA/μm²) and **b** temperature gradient (unit: °C/μm) of the waffle layout in Fig. 3.18, titled view from the back. The *circles* indicate the problematic region

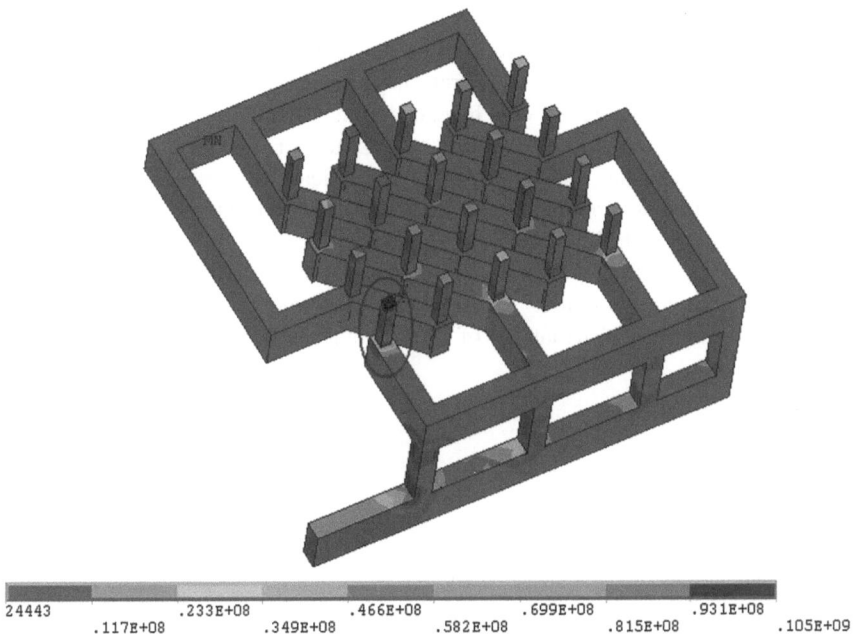

24443   .233E+08   .466E+08   .699E+08   .931E+08
    .117E+08   .349E+08   .582E+08   .815E+08   .105E+09

Fig. 3.20 Total AFD distribution (unit: atoms/μm^3·s) of the waffle layout in Fig. 3.18, titled view from the back. The *circle* indicates the problematic region

3.6 Summary

In this chapter, 3D circuit model of the output stage of a class-AB amplifier was compared with the line-via test structure, under both the EM test condition and the circuit operation condition. It was observed that though the maximum AFD locations of the circuit model coincided with that of the line-via test structure at EM test temperature, the locations for the two structures are different at the circuit operation temperature as current density was no longer the dominant driving force for EM at this temperature. The EM modeling based on the line-via test structure might not be able to identify the correct EM failure sites for a circuit at its operation temperature. This study again implied the need for the complete 3D circuit modeling in the EM study.

A few sets of models were constructed to compare the difference in the values and locations of the maximum total AFD when there was a change in interconnect structure. Higher maximum total AFD was observed for the structures with a longer via and contact distance, shorter inter-transistor distance, smaller number of contacts, and more metal layers. Some changes in the interconnect structure could cause a change in the maximum total AFD location. The observations were consistent with the experimental results and thus validated the capability of the 3D finite element circuit model for performing the EM lifetime comparison when there was a structural change in the interconnects.

In the case of small transistors, there were fewer contacts on the source and drain interconnects for the transistor with more fingers due to the limitation in the metal line length of the source and drain interconnects. An increase in the maximum total AFD was observed with increasing finger numbers, and the value of the maximum total AFD depended on the smaller number of contacts on the source and drain interconnects. The impact was insignificant in the case of large transistors as the number of contacts on the source and drain interconnects is similar for the transistors with different finger numbers.

The use of the waffle layout will degrade the EM reliability of the circuit due to the non-uniform current density and temperature gradient distributions as a result of the non-uniform distribution of the contacts on the source and drain interconnects.

References

1. Baker RJ (2004) CMOS circuit design layout and simulation, Revised 2nd edn. IEEE Press Series on microelectronic Systems, Wiley, NY
2. He F, Tan CM (2012) Comparison of electromigration simulation in test structure and actual circuit. Appl Math Model 36:4908–4917
3. Lin M, Jou N, Liang JW, Su KC (2009) Effect of multiple via layout on electromigration performance and current density distribution in copper interconnect. In: IEEE International Reliability Physics Symposium, 2009, pp 844–847
4. Lin MH, Lin YL, Chang KP, Su KC, Wang T (2006) Copper interconnect electromigration behavior in various structures and precise bimodal fitting. Jap J Appl Phys Part 1 45(2A):700–709
5. Chen CT, Hsu T-S, Jeng R-J, Yeh H-C (2000) Enhancing the glass-transition temperature of polyimide copolymers containing 2,2-bipyridine units by the coordination of nickel malenonitriledithiolate. J Polymer Sci, Part A: Polymer Chem 38(3):498–503
6. Meeker WQ, Escobar LA (1998) Statistical methods for reliability data. Wiley, New York
7. Suo Z (2003) Reliability of interconnect structures. Interfacial and Nanoscale Failure. Comprehensive Structural Integrity, vol 8. p 265
8. He F, Tan CM (2012) 3D simulation-based research on the effect of interconnect structures on circuit EM reliability. World J Model Simul 8:271–284
9. Tan CM, Roy A (2007) Electromigration in ULSI interconnects. Materials Sci Eng R 58(1–2):1–75
10. Li W, Tan CM, Raghavan N (2009) Dynamic simulation of void nucleation during electromigration in narrow integrated circuit interconnects. J Appl Phys 105(1):014305
11. Rzepka S, Banerjee K, Meusel E, Hu CM (1998) Characterization of self-heating in advanced VLSI interconnect lines based on thermal finite element simulation. IEEE Trans. Components, Packaging, and Manufacturing Technology, Part A 21(3):406–411
12. He F, Tan CM (2012) Effect of IC layout on the reliability of CMOS amplifiers. Microelectron Reliab 52:1575–1580
13. Layout of analog CMOS integrated circuit, http://ims.unipv.it/Microelettronica/Layout02.pdf

Chapter 4
Interconnect EM Reliability Modeling at Circuit Layout Level

4.1 Introduction

The circuit models built in Chaps. 2 and 3 only consider the intra-block interconnects and simple inter-block connections up to Metal 2. Realistic circuits consist of a large number of transistors and other circuit components connected by complex inter-block connections made of multiple metal layers. In this chapter, a complete 3D circuit model including both intra- and inter-block interconnects is constructed. Electro-thermo-structural simulations are performed, and the modifications that can help enhancing the EM reliability of the circuit are carried out based on the observations in the simulation.

4.2 Model Construction and Simulation Setup

Low noise amplifier (LNA) is widely used in communication and microwave system. It boosts the desired signal power with little noise and distortion. The LNA circuit is chosen as an example in this chapter because it is a simple and common circuit and has wide application. Also, the LNA is susceptible to reliability degradation because it is very sensitive to circuit condition. The construction and modeling procedures introduced in this section are applicable to other circuits as well.

The schematic and layout of a two-stage LNA circuit is shown in Figs. 4.1 and 4.2, respectively [1, 2]. The first stage of the circuit consists of a main input transistor M1 and a cascode transistor M2, and the second stage consists of a main input transistor M3 and a cascode transistor M4. The supply voltage from V_{dd} and the input voltage V_{in} are both 1.8 V, and the bias voltage V_{bias1} and V_{bias2} are both 0.7 V, as indicated in Fig. 4.1. The operating frequency of the circuit is 5.8 GHz.

Figure 4.3 is the partial side view of Transistor M2 along AA' in Fig. 4.2b, showing the source and drain interconnects of the model. The gate interconnect is not shown for clarity. The current flows vertically along the intra-block

C. M. Tan and F. He, *Electromigration Modeling at Circuit Layout Level*,
SpringerBriefs in Reliability, DOI: 10.1007/978-981-4451-21-5_4, © The Author(s) 2013

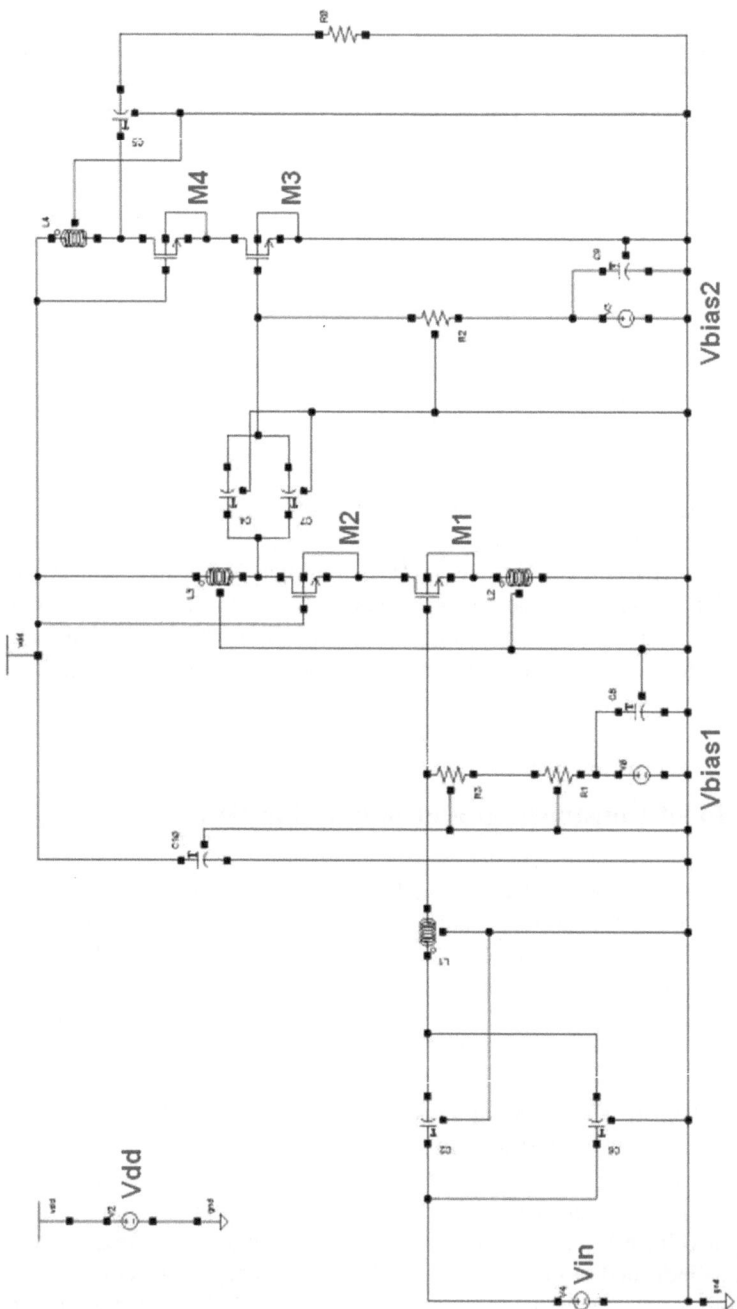

Fig. 4.1 Schematic of a simple LNA circuit (Reprinted from Microelectronics [1], Copyright (2012), with permission from Elsevier)

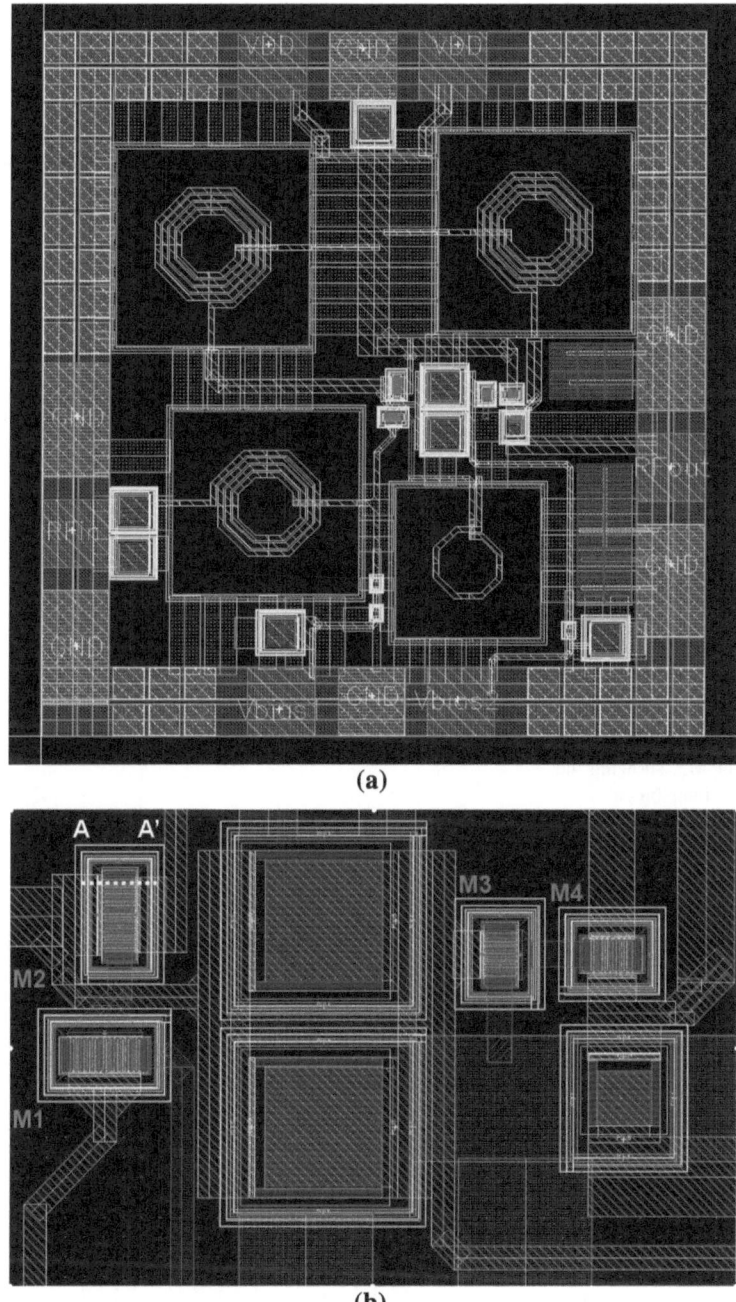

(a)

(b)

Fig. 4.2   **a** Layout of the circuit in Fig. 4.1 and **b** the zoom-in view at the transistor region (Reprinted from He and Tan [1], Copyright (2012), with permission from Elsevier)

interconnect to/from the source and drain regions, and horizontally along the inter-block interconnect (Metal 3 and 6 in this case) to/from other circuit components. The source and drain currents extracted from Cadence are used as the inputs in the ANSYS simulation, with the assumption that the same amount of current flows in the top and bottom of the metal layers on the same source or drain interconnect (i.e., the current flows in the source interconnect at location A and B and in the drain interconnect at location C and D are assumed to be the same). There are changes in the current density along the current flow paths due to the variation in the track and via widths and the metal thicknesses. Current crowding and high thermo-mechanical stress built up occur at the locations with significant dimensional or geometrical changes, such as the contact/metal and via/metal interfaces, as indicated by the rectangular boxes in Fig. 4.3.

As our work is focusing on the EM reliability of the interconnects, the circuit components (i.e., transistors, capacitors, inductors, and resistors) are assumed to be reliable. The study in [3] showed that the EM lifetime is independent of the surrounding dummy structures for metal line width larger than 0.063 μm. The LNA circuit under study is using 0.18 μm technology, and the minimum metal line width is 0.23 μm. Therefore, the dummy structures in this circuit do not affect the EM lifetime of the model, and they are removed during the 3D model construction in order to reduce the simulation time and computation memory.

Fig. 4.3 Partial side view of Transistor M2, showing the intra- and inter-block connections at the source and drain regions. The *arrows* indicate the direction of the current flow (Reprinted from He and Tan [1], Copyright (2012), with permission from Elsevier)

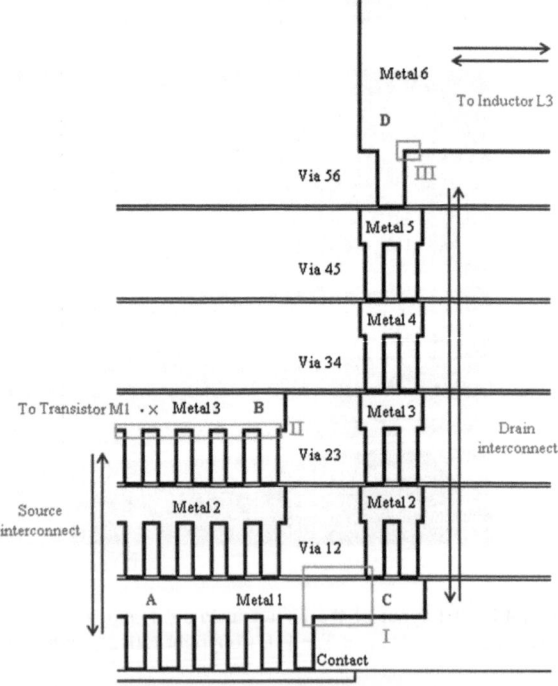

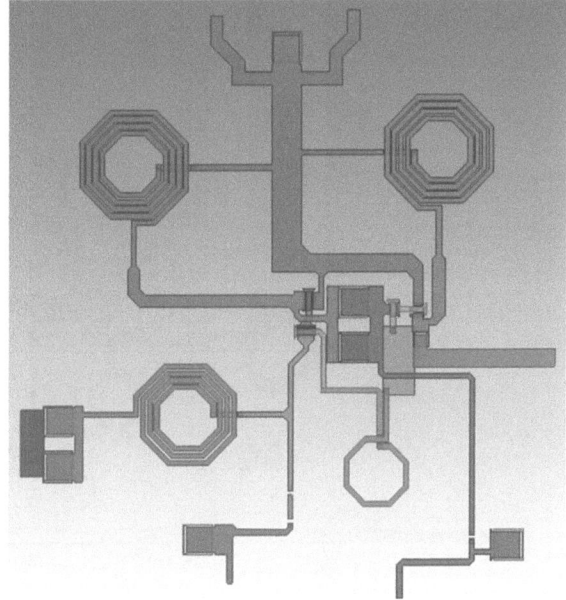

Fig. 4.4 Top view of the full model of the LNA circuit after removing the dummy structures, showing only the interconnects (Reprinted from He and Tan [1], Copyright (2012), with permission from Elsevier)

The simulation is first conducted on a coarsely meshed full model that includes all the inter-block interconnects, as shown in Fig. 4.4. The intra-block interconnects are lumped together as simple blocks. Skin effect, reflected by a 10 % increase in resistance below the skin depth region (i.e., 0.87 µm at the operating frequency of 5.8 GHz in this case), is included in the model [4]. Here, only the thick metal top, Metal 6 with a thickness of 2.34 µm, is affected by the skin effect.

The operating temperature of a LNA with a supply voltage of 1.8 V is between −40 and 85 °C [5, 6]. Here, the worst-case temperature of 85 °C is used. Quadratic tetrahedral element SOLID 98 is used in the ANSYS simulation, and the model construction and the simulation setup are the same as in Chaps. 2 and 3.

4.3 Distributions of Atomic Flux Divergences

4.3.1 Total AFD Distribution of the Full Model

The total AFD distribution of the full model at the end of one load cycle is shown in Fig. 4.5. The location of the maximum total AFD is found at the turning corner of the source interconnect of Transistor M1, as indicated by the arrow in Fig. 4.5. As the locations of the maximum current density, temperature and stress gradient do not change after many load cycles, this location can be treated as the final maximum total AFD location of the circuit.

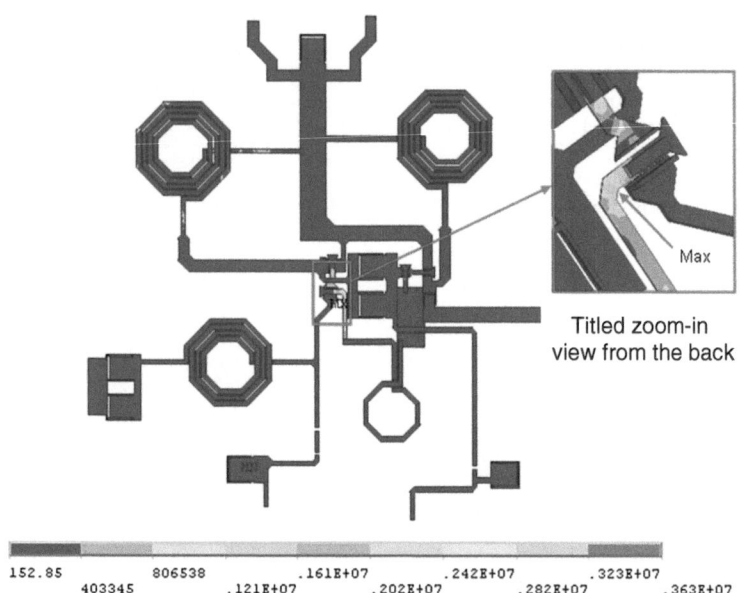

Titled zoom-in
view from the back

```
152.85        806538      .161E+07        .242E+07        .323E+07
       403345        .121E+07      .202E+07        .282E+07      .363E+07
```

Fig. 4.5 Total AFD distribution of the full model of the LNA circuit (unit: atoms/$\mu m^3 \cdot$s). The *arrow* indicates the maximum total AFD location (Reprinted from He and Tan [1], Copyright (2012), with permission from Elsevier)

The current density and temperature gradient distributions of the full mode are shown in Figs. 4.6 and 4.7, respectively. In the circuit layout, the V_{dd} lines are connected to the gate of Transistors M2 and M4. With negligible gate current, the current density and the resulting temperature gradient of the V_{dd} lines are small (they are all inside the blue regions as can be seen in Figs. 4.6 and 4.7). These interconnects are therefore of little concern in this work.

Although the current flows in the V_{dd} and ground lines connected to the inductors are high, these lines are made of the thick top metal (Metal 6 in this case). The maximum current density in these interconnects is more than 10 times smaller than that in the source and drain interconnects at the transistor region which are made of the thinner Metal 3 layer. Table 4.1 shows the metal layer type and shape as well as the line width of the source and drain interconnects at the transistor regions. The highest current density is found at the source interconnect of Transistor M1 with a sharp turn, as indicated by the arrow in Fig. 4.6.

The interconnect temperature is affected by the heat convection to the surrounding through the outer surface of the chip, the heat conduction through the heat sink, and Joule heating due to current flow in the interconnects, in which the latter two are more dominant. If the Joule heating effect is ignored, the upper metal layers are found to have a higher temperature than the lower metal layers due to a larger distance between the upper layers and the heat sink [7]. However, when the Joule heating effect is taken into consideration, the heat flux in the current carrying

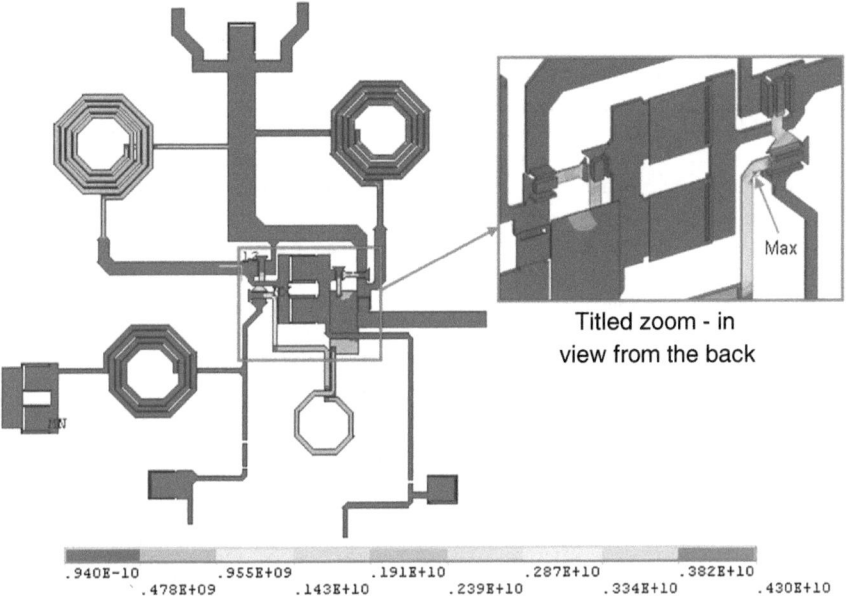

.940E-10   .955E+09   .191E+10   .287E+10   .382E+10
   .478E+09   .143E+10   .239E+10   .334E+10   .430E+10

Fig. 4.6 Current density distribution of the full model (unit: pA/μm^2). The *arrow* indicates the location with the highest current density

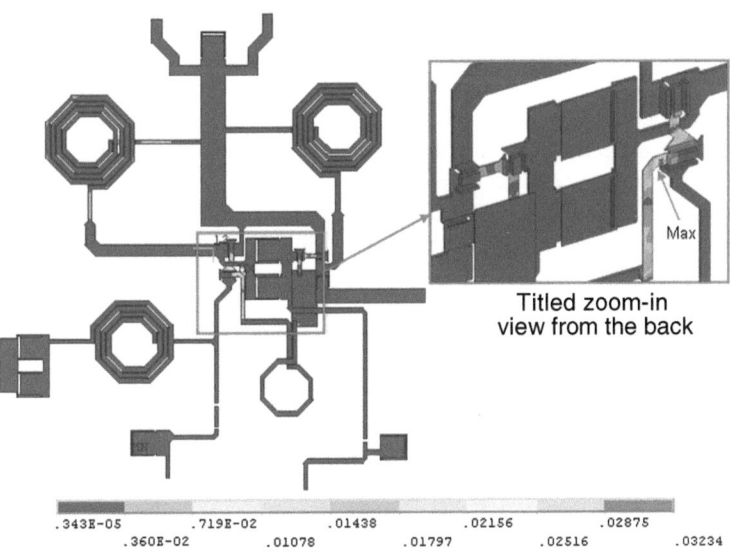

.343E-05   .719E-02   .01438   .02156   .02875
   .360E-02   .01078   .01797   .02516   .03234

Fig. 4.7 Temperature gradient distribution of the full model (unit: °C/μm). The *arrow* indicates the location with the highest temperature gradient

Table 4.1 Metal layer type, shape and line width of the source and drain interconnects of the four transistors (Reprinted from Microelectronics [2], Copyright (2012), with permission from Elsevier)

Interconnect		Type	Shape	Line width (μm)
Transistor M1	Source	Metal 3	Sharp turn	6.40
	Drain	Metal 3	Small turn	6.40
Transistor M2	Source	Metal 3	Small turn	6.40
	Drain	Metal 6	Simple strip	15
Transistor M3	Source	Metal 3	Simple strip	6.40
	Drain	Metal 3	Simple strip	6.40
Transistor M4	Source	Metal 3	Simple strip	6.40
	Drain	Metal 6	Sharp turn	12

lines becomes the determining factor for the temperature and temperature gradient of the interconnects [8]. The temperature rise in a single line interconnect is proportional to the square of the current density [9], and it increases with increasing current density, so does the temperature gradient [10].

Local high current density causes a high temperature gradient at the source and drain interconnects of Metal 3, especially at the sharp turning corners, as shown by the orange and the red regions in Fig. 4.7. Based on our simulation results in Chaps. 2 and 3, the total AFD under the circuit operation condition is mainly determined by AFD_S, which in turn depends on the product of the local thermo-mechanical stress gradient and the temperature gradient. The source interconnect of Transistor M1 with a sharp turn therefore has the highest AFD_S and the highest total AFD due to the highest temperature gradient at this location.

However, lumping the intra-block interconnects together causes a sharp drop in the thermo-mechanical stress gradient due to the uniformity in structure in the simplified blocks. This results in a drop in the values of AFD_S and hence the total AFD. This is especially true for the transistor with more fingers (i.e., more complicated structure), as can be seen in Fig. 4.8.

Moreover, the over-simplification of the interconnect structure causes the disappearance of the structures with significant dimensional or geometrical changes.

Fig. 4.8 Percentage drop in the maximum total AFD due to structural simplification versus the transistor finger number (Reprinted from He and Tan [1], Copyright (2012), with permission from Elsevier)

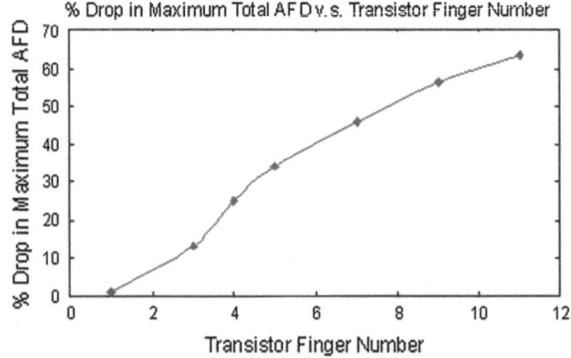

Comparing Fig. 4.3 (before simplification) and Fig. 4.9 (after simplification), we can see that the complex structures at Region II disappear. The disappearance of the complex structures may affect the current density and thermo-mechanical stress distributions of the model and thus results in an inaccurate prediction. The coarse mesh of the full model also reduces the accuracy of the simulation results further.

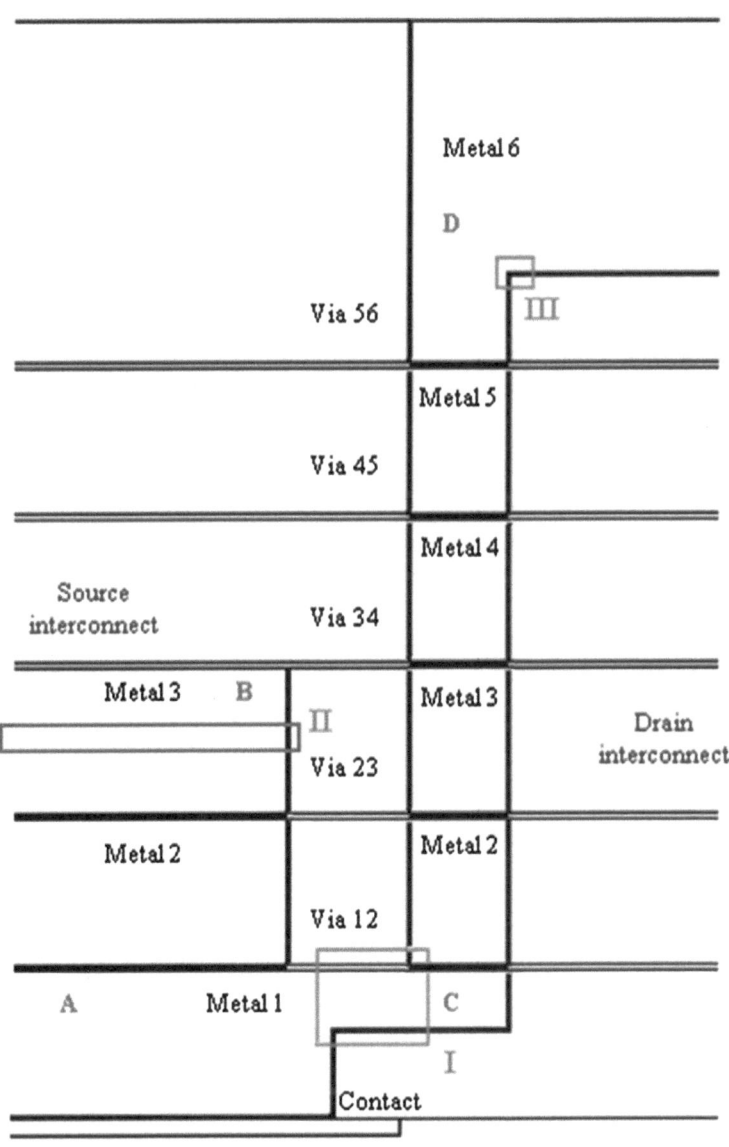

Fig. 4.9   Partial side view of Transistor M2 after structural simplification

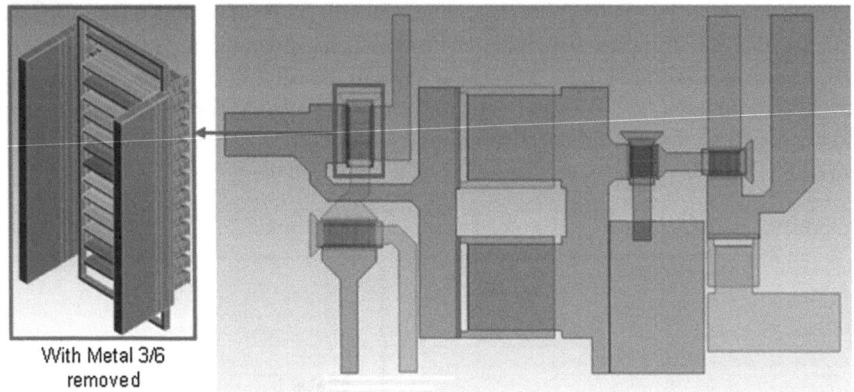

With Metal 3/6
removed

Fig. 4.10 Sub-modeling at the transistor region, with the inclusion of the detailed intra-block interconnect structures. The surrounding materials are removed for clarity (Reprinted from He and Tan [1], Copyright (2012), with permission from Elsevier)

Therefore, the simulation is performed again on a finely meshed sub-model at the "problematic" region with the inclusion of the detailed structures of the intra-block interconnects. Due to the limitation in computation memory in our case, the model file size cannot be bigger than 100 MB. The model size limit is exceeded if all the intra-block contacts and vias are included, and hence, a simplification is performed to reduce the model size. The simulation in [11] showed that the temperature difference of the lumped structure (i.e., the contacts and vias on the same metal line are lumped together) and the original structure is less than 5 %, mostly due to the inherent error. Hence, the lumped contact and via structure can be used. The structural simplification at the contact and via regions is shown in the left figure of Fig. 4.10. The simulation results of the sub-model are discussed in the next section.

4.3.2 Total AFD Distribution of the Sub-Model

The maximum total AFD of the sub-model at the end of one load cycle is also at the source interconnect of Transistor M1 as shown in Fig. 4.11, which corresponds to the simulation results of the full model in Fig. 4.5. An increase in the value of the total AFD is observed for the sub-model when compared with that of the full model. This is due to the increase in the thermo-mechanical stress gradient after the inclusion of the detailed structures of the intra-block interconnects.

The total AFD distribution and the distributions of the AFDs due to the three driving forces of the sub-model at Transistor M1 region (i.e., the region where EM is the most severe) are shown in Fig. 4.12, and the distributions of the current

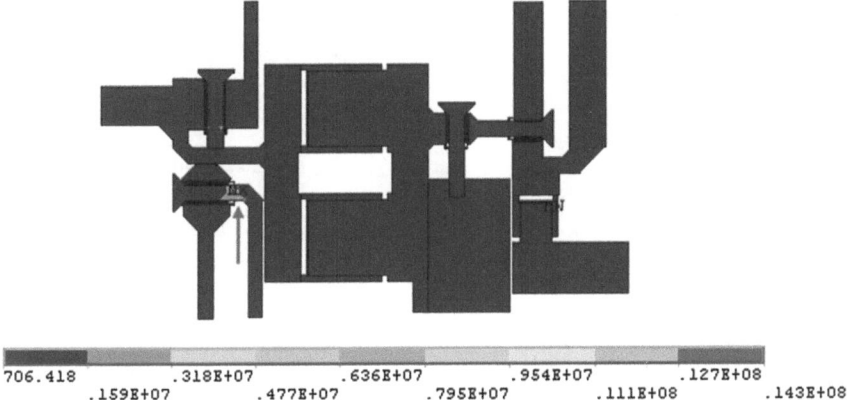

706.418   .318E+07   .636E+07   .954E+07   .127E+08
    .159E+07   .477E+07   .795E+07   .111E+08   .143E+08

Fig. 4.11 Total AFD distribution of the sub-model of the LNA circuit (unit: atoms/μm^3·s). The *arrow* indicates the maximum total AFD location (Reprinted from He and Tan [1], Copyright (2012), with permission from Elsevier)

density, temperature gradient, and thermo-mechanical stress gradient are shown in Fig. 4.13.

As explained previously, the total AFD distribution follows that of AFD_S, and this agrees with the observations in Fig. 4.12a and c. In addition, it is observed from Fig. 4.12a that, besides the sharp turn of the source interconnect of Transistor M1, the interface between Via 23 and Metal 3 (i.e., Region II in Fig. 4.3) is also a location with high total AFD values (i.e., above 6.36×10^6 atoms/μm^3·s). Although the current density and the temperature gradient at the Via 23/Metal 3 interface are around half of that at the sharp turn of the source interconnect, the 90° turning of the interconnect structure at this interface builds up significant thermo-mechanical stress, as shown in Fig. 4.13c. The thermo-mechanical stress gradient at this region is almost 2 times larger than that at the sharp turn of the source interconnect. Therefore, the Via 23/Metal 3 interface has comparably high AFD_S and total AFD values as that at the sharp turn of the source interconnect.

The Via 56/Metal 6 interface (i.e., Region III in Fig. 4.3) is another location with this kind of high stress and stress gradient built up. However, as explained in Sect. 4.3.1, the current density and temperature gradient at the thick metal top are more than 10 times smaller than those at the thin Metal 3 layer. AFD_S and the total AFD at this location are also more than 10 times smaller than those at the two aforementioned locations. Therefore, this region is not a major concern.

From Fig. 4.13a, we can see that besides the sharp turn of the source interconnect of Transistor M1, the contact/Metal 1 and Via 12/Metal 2 interfaces of the drain interconnect of Transistor M1 (i.e., Region I in Fig. 4.3) is also a current crowding site, and it has the highest current density, but the presence of the large amount of contacts and vias separated at a distance smaller than the thermal diffusion length (i.e., the distance at which the heat flux reduces e times from its surface value [12], usually 1–20 μm) reduces the temperature and temperature

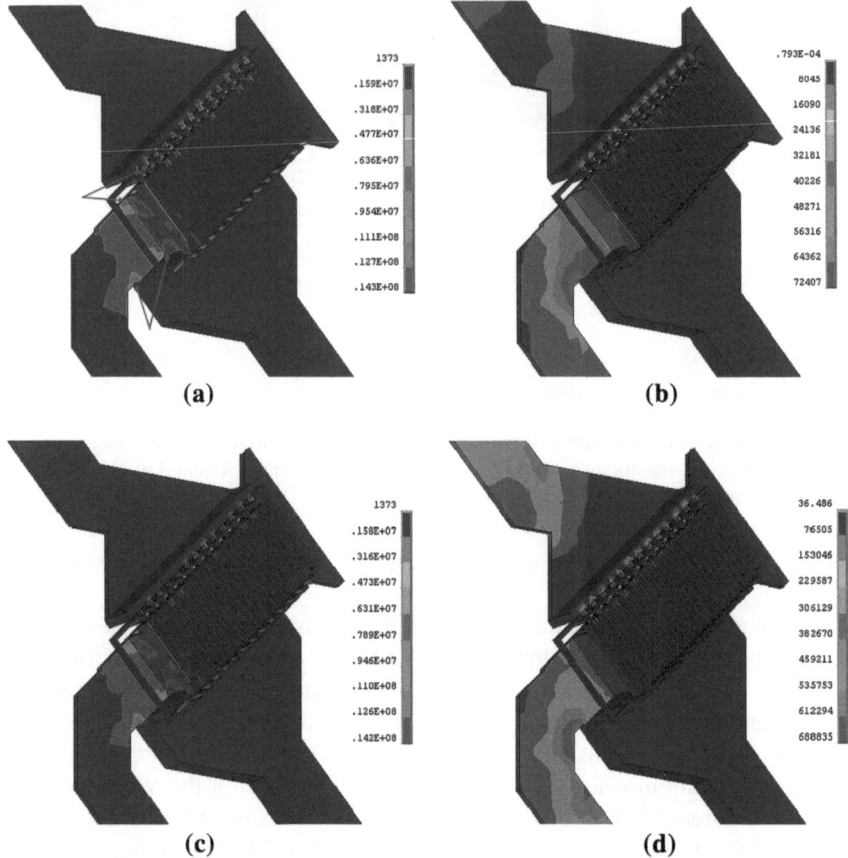

Fig. 4.12 AFD distributions due to (**a**) total AFD (**b**) electron wind force (**c**) thermo-mechanical stress gradient–induced driving force and (**d**) temperature gradient–induced driving force (unit: atoms/μm³·s) of the sub-model at Transistor M1 region, titled view from the back. The *arrows* indicate the locations with high total AFD (Reprinted from He and Tan [1], Copyright (2012), with permission from Elsevier)

gradient at Metal 1 [13, 14]. As a result, the maximum AFD_S and total AFD here are around 10 % smaller than those at the sharp turn and the Via 23/Metal 3 interface, but the values are still higher than that at other places.

For Transistor M3, the source and drain interconnects are in simple rectangular shape. The regularity in shape greatly reduces the current density and the temperature gradient in Metal 3. The elimination of the sharp turn in the interconnects eliminates the possible EM weak spot at this location.

On the other hand, the number of contacts and vias at the Transistors M3 and M4 regions is only around 2/3 of that at the Transistors M1 and M2 regions due to their smaller transistor sizes. The current crowding at Region I and the reduced number of dummy contacts and vias cause a large rise in temperature. The location

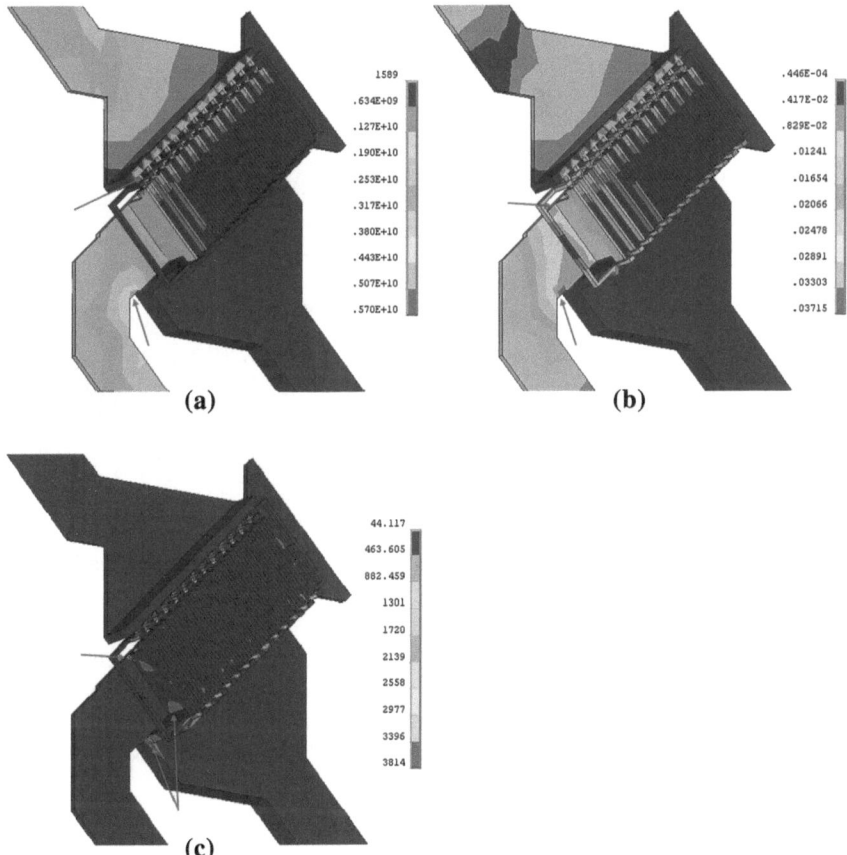

Fig. 4.13 Distributions of (**a**) current density (unit: pA/μm²) (**b**) temperature gradient (unit: °C/μm) and (**c**) thermo-mechanical stress gradient (unit: MPa/μm) of the sub-model at the Transistor M1 region, titled view from the back. The *arrows* indicate the locations with high values

of the maximum temperature gradient coincides with that of the maximum current density for Transistor M3. The effect of temperature gradient on AFD_S takes over that of the thermo-mechanical stress gradient, and thus, Region I becomes the place with the maximum AFD_S and total AFD, as shown in Fig. 4.14. Region II has high thermo-mechanical stress gradient of more than 3,500 MPa/μm, and the location of the 2nd highest AFD_S and total AFD is found at Region II.

In summary, this section discussed the possible EM failure sites of the LNA circuit model. It is found that the full model is able to identify the most server EM failure location. However, some possible failure locations are overlooked in the full model due to the oversimplification in the interconnect structures. This problem is remediated by using the sub-model that includes the detailed

Fig. 4.14 Total AFD
distribution of the sub-model
at Transistor M3 region (unit:
atoms/μm$^3\cdot$s), titled view
from the back. The *arrow*
indicates the maximum total
AFD location (Reprinted
from He and Tan [1],
Copyright (2012), with
permission from Elsevier)

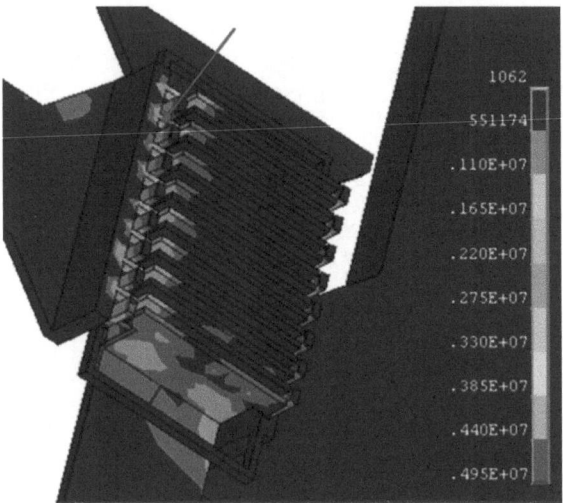

interconnect structures at the transistor region. Additional EM failure locations due
to high thermo-mechanical stress gradient are identified in the sub-model
simulation.

The total AFD distribution of the circuit under the circuit operation condition is
the combined effect of the three driving forces, and is not purely due to current
density as one usually assumes. In fact, the location with the highest current
density may (e.g., Transistor M3) or may not (e.g., Transistor M1) be the location
of the maximum total AFD.

4.4 Effects of Layout and Process Parameters on Circuit EM Reliability

As discussed in Sect. 2.3.4, the total AFD of the circuit is inversely proportional to
the EM lifetime of the circuit. To improve the EM reliability of the circuit, the
total AFD should be reduced. With the findings in the previous section, there are
two possible ways for the reduction in the total AFD:

1. Reduce the temperature gradient by

   - Reducing the current density (e.g., increase the interconnect line width).
   - Relieving the degree of current crowding (e.g., use a smoother edge or a
     simple strip instead of the structure with a sharp turn).
   - Reducing the temperature of the circuit (e.g., increase the inter-transistor
     distance)

2. Reduce the thermo-mechanical stress gradient by

- Avoiding significant geometry change.
- Reducing the intrinsic stress in the interconnects (e.g., decrease the SFT of the metallization).

This section shows some common practices that are usually adopted for improving the EM reliability of an IC. The LNA model is used as an example.

4.4.1 Line Width and Degree of Turning

The high total AFD at the sharp turn of Transistor M1 is due to the high temperature gradient at this location as a result of current crowding. It can be reduced

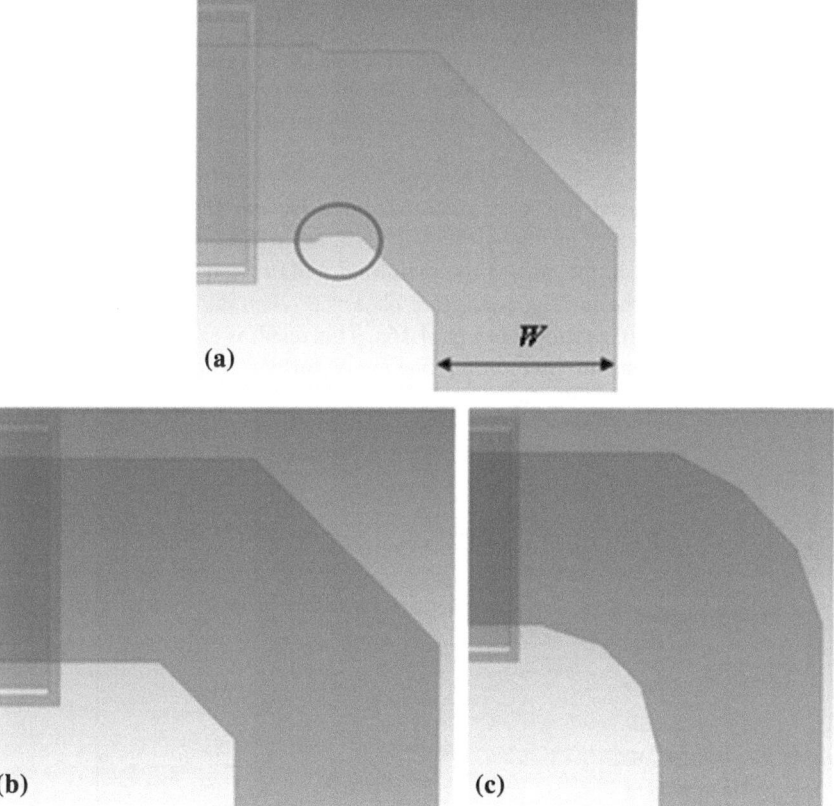

Fig. 4.15 Layout modification for the EM lifetime improvement at Transistor M1 region. (a) The original layout with a high current density at the circled region, line width $W = 6.00$ μm (b) the modified layout that removes the sharp turn at the circled region, $W = 6.40$ μm and (c) the modified layout with a smoother edge, $W = 6.40$ μm (Reprinted from He and Tan [1], Copyright (2012), with permission from Elsevier)

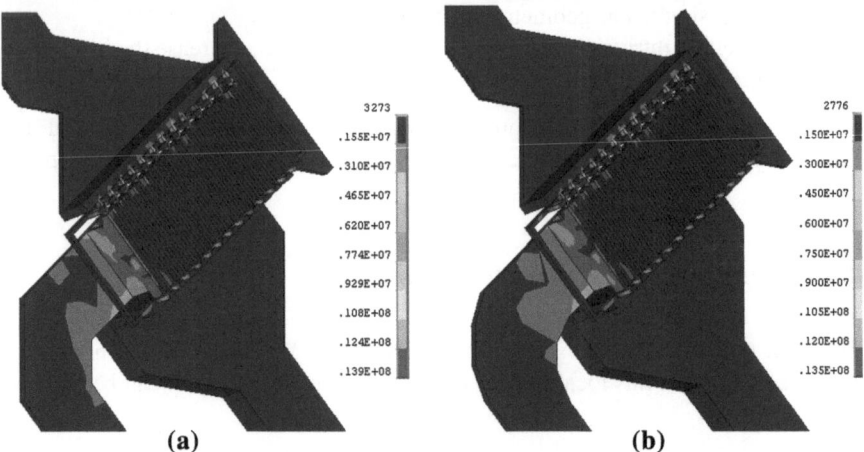

(a)   (b)

Fig. 4.16 Total AFD distributions of the modified layout at Transistor M1 region with (**a**) the sharp turn removal by using a wider metal line and (**b**) the use of a smoother edge (unit: atoms/ μm^3·s), titled view from the back (Reprinted from He and Tan [1], Copyright (2012), with permission from Elsevier)

by using a wider metal line or a smoother edge for the source interconnect of Transistor M1 as shown in Fig. 4.15. A 2.80 % reduction in the maximum total AFD and a decrease in the area of the high total AFD region (i.e., the region with light blue color) at the turning corner are observed when the line width increases from 6.00 to 6.40 μm, as shown in Fig. 4.16a. This result is expected, and it agrees with the observations in literature [15–17]. A further reduction of 2.80 % is observed in Fig. 4.16b with the use of a smoother edge.

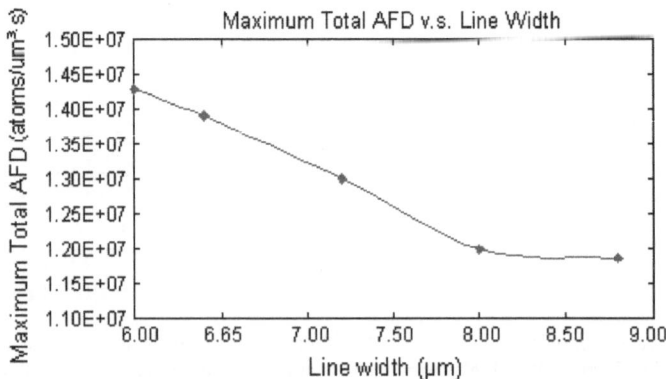

Fig. 4.17 Maximum total AFD vs the line width of the source interconnect of Transistor M1 (Reprinted from He and Tan [1], Copyright (2012), with permission from Elsevier)

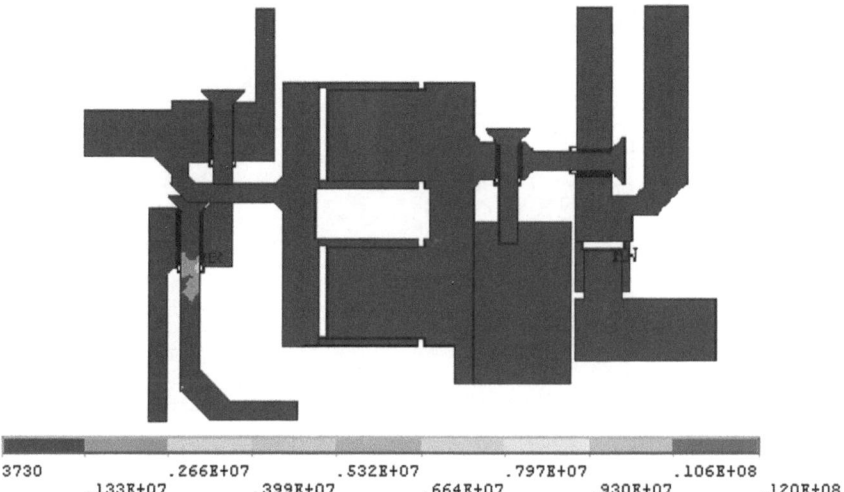

```
3730        .266E+07        .532E+07        .797E+07        .106E+08
     .133E+07        .399E+07        .664E+07        .930E+07        .120E+08
```

Fig. 4.18   Total AFD distribution of the sub-model with a different orientation for Transistor M1 (unit: atoms/μm^3·s) (Reprinted from He and Tan [1], Copyright (2012), with permission from Elsevier)

Figure 4.17 shows the decrease in the maximum total AFD with increasing line width. Up to 16.78 % reduction in the maximum total AFD and 20.17 % improvement in the EM lifetime can be achieved when the line width increases to 8.00 μm. There is no further change at above 8.00 μm.

4.4.2   Transistor Orientation

Figure 4.18 shows a 16.09 % reduction in the value of the maximum total AFD at the source interconnect of Transistor M1 after a 90° rotation of the transistor. The change in shape of the source interconnect eliminates the high current crowding at the turning corner and thus effectively reduces the temperature gradient and the maximum total AFD of the circuit. However, this rotation produces another sharp turn at the inter-transistor interconnect between Transistors M1 and M2, and the improvement in lifetime may not always be possible and is case dependent.

Figure 4.19 is another modified layout with the transistor finger number doubled. The presence of the two inductors around Transistor M1 forces a 90° rotation of M1, and this layout has the same transistor orientation as in Fig. 4.18. The total AFD distribution of this layout is shown in Fig. 4.20.

As observed from Fig. 4.20, there is a significant increase in the value of the maximum total AFD for this layout version when compared with the original layout and the location of the maximum total AFD moves from Transistors M1 to M2. Different from the simple two metal layer circuit models in Chaps. 2 and 3,

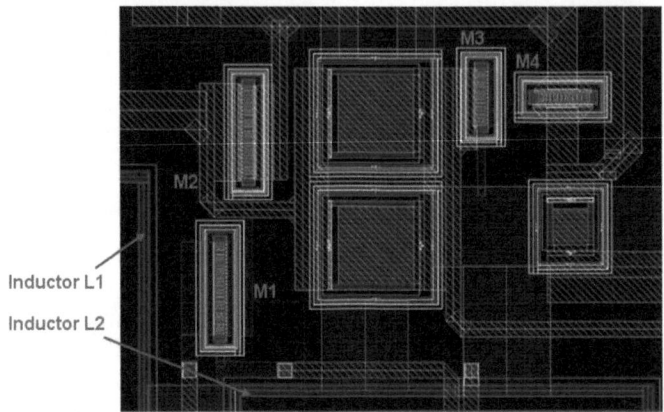

Fig. 4.19 Modified layout with the transistor finger number doubled. There is a limitation in space at Transistor M1 region (Reprinted from [2], Copyright (2012), with permission from Elsevier)

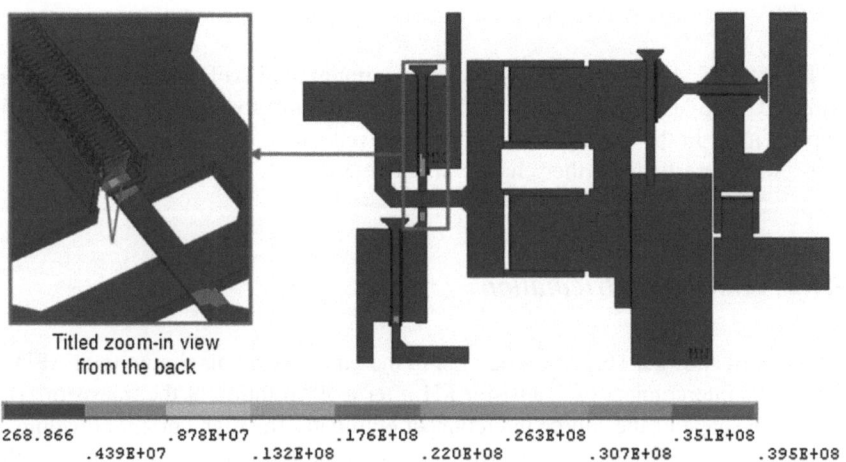

Fig. 4.20 Total AFD distribution of the sub-model with the transistor finger number doubled (unit: atoms/$\mu m^3 \cdot s$). The *arrows* indicate the locations with high total AFD (Reprinted from [2], Copyright (2012), with permission from Elsevier)

the source of the transistors of the LNA circuit is connected to the inter-block interconnects via the contacts, Metal 1, Via 12, Metal 2, Via 23, and Metal 3, as can be seen from the tilted back view of the transistor in Fig. 4.20. The width of the source interconnect made of Metal 3 is determined by the width per finger of the transistor. When the number of fingers is doubled, the width per finger of the transistor and thus the line width of the source interconnect become half of that of the layout shown in Fig. 4.2. As a result, very high current crowding is observed at the source interconnect of Transistors M1 and M2 because of the narrower Metal 3 line. The rotation of Transistor M1 eliminates the current crowding at the turning corner of the source interconnect of Transistor M1, but the turning corner of the inter-transistor interconnect between Transistors M1 and M2 becomes another region with high total AFD (i.e., the region with light blue color) due to its small line width. The location of the maximum total AFD is found at the narrow and long source interconnect of Transistor M2 with the highest temperature gradient. The temperature gradient of the circuit is almost doubled with the halved inter-block interconnect line width; 176.22 % increase in the value of the maximum total AFD is observed, and the EM lifetime for this layout is expected to be less than half of that of the layout shown in Fig. 4.2.

4.4.3 Inter-Transistor Distance

Changing the placement of the electronic components is a common practice for the modification of the temperature and stress distributions in printed wiring board (PWB) and multichip module (MCM) [18–20]. This measure is attempted in this circuit model.

In the LNA circuit, all the four transistors are the main heat generating sources, and their placements do have an impact on the circuit temperature. Figure 4.21 shows two modified layouts of the LNA circuit with different inter-transistor distance. The inter-transistor distances of the original and the modified layout designs are shown in Table 4.2.

The total AFD distributions of the modified layouts are shown in Fig. 4.22. The increase in the inter-transistor distance reduces the value of the maximum total AFD with little change in the location of the maximum total AFD.

Table 4.2   The inter-transistor distance of the LNA circuit with different layout designs

Horizontal distance (μm)	Layout 4.2	Layout 4.21(a)	Layout 4.21(b)
M1 to M2	7.69	65.10	0.00
M1 to M4	132.425	218.425	34.925
M3 to M2	99.50	209.50	34.925
M3 to M4	29.175	74.125	0.00
Vertical distance (μm)			
M1 to M2	35.25	89.70	32.75
M1 to M4	25.40	97.35	29.175
M3 to M2	9.875	18.125	32.75
M3 to M4	2.175	25.675	29.175
Center-to-center distance (μm)			
M1 to M2	36.08	110.83	32.75
M1 to M4	134.84	239.14	45.51
M3 to M2	99.99	210.28	47.88
M3 to M4	29.26	78.45	29.175

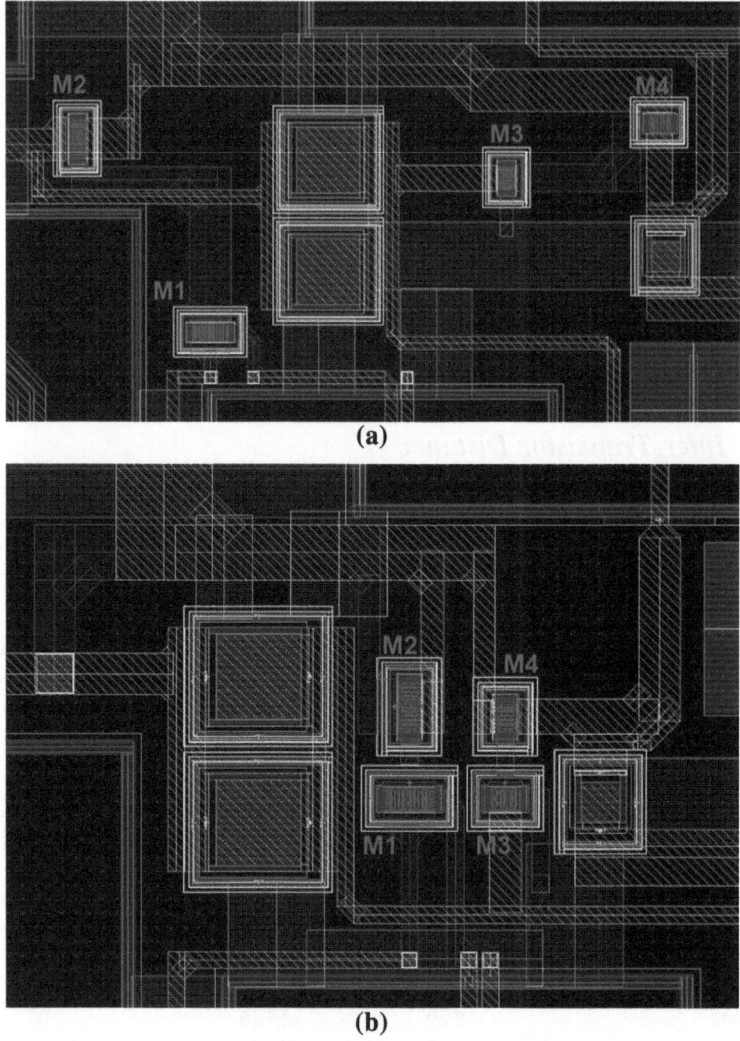

Fig. 4.21 Zoom-in view at the transistor region of the modified layouts with (**a**) the transistors placed further apart ($\sim 2x$) and (**b**) the transistors placed closer together (the minimum distance) (Reprinted from [2], Copyright (2012), with permission from Elsevier)

The current density and thermo-mechanical stress gradient for the layouts shown in Figs. 4.2, 4.21a and b are similar due to the same current and voltage loads and similar transistor architecture. As a result, the total AFD distributions of these three layouts are similar. The regions with high total AFD are found at the sharp turn of the source interconnect of Transistor M1, the Via 23/Metal 3 interface, and the contact/Metal 1/Via 12 interfaces of the drain interconnect, as explained in Sect. 4.3.2.

The impact of the inter-transistor distance on the total AFD of the models in Sect. 3.4 is insignificant as there is only one main heat generating transistor for that circuit. However, the inter-transistor distance does have an impact on the reliability of the LNA. In the LNA circuit, all the four transistors have significant amount of heat generation due to high current flows inside the transistors. The increase in the inter-transistor distance in Layout 4.21(a) effectively reduces the temperature and temperature gradient of the circuit, resulting in 49.23 % reduction in the maximum total AFD when compared with the original layout.

As the "problematic" region consists of Transistors M1 and M2 as can be seen in Fig. 4.22, the simulation is re-done with different inter-transistor distances between Transistors M1 and M2, and the drop in the maximum total AFD with

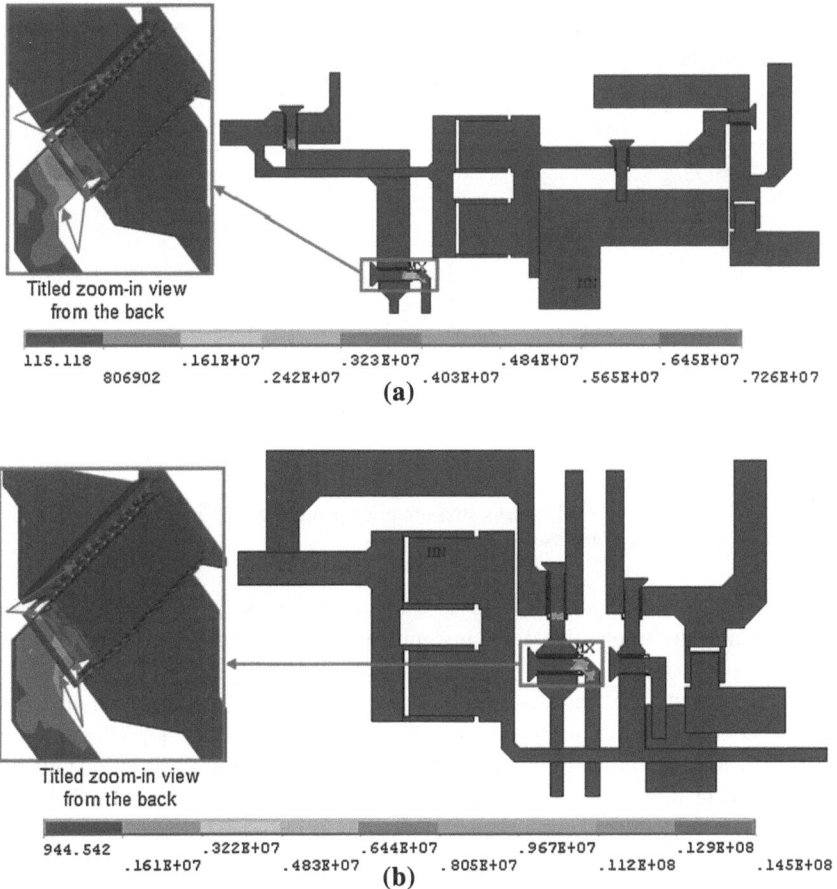

Fig. 4.22 Total AFD distributions of the modified layouts with (**a**) the transistors placed further apart and (**b**) the transistors placed closer together (unit: atoms/$\mu m^3 \cdot s$). The *arrows* indicate the locations with high total AFD (Reprinted from [2], Copyright (2012), with permission from Elsevier)

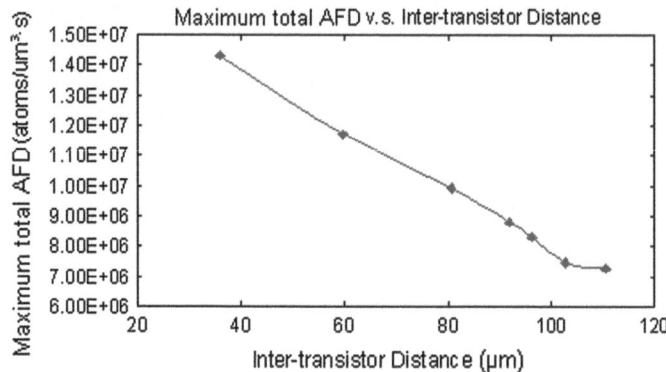

Fig. 4.23   Drop in the maximum total AFD with increasing inter-transistor distance (center-to-center) between Transistors M1 and M2 (Reprinted from [2], Copyright (2012), with permission from Elsevier)

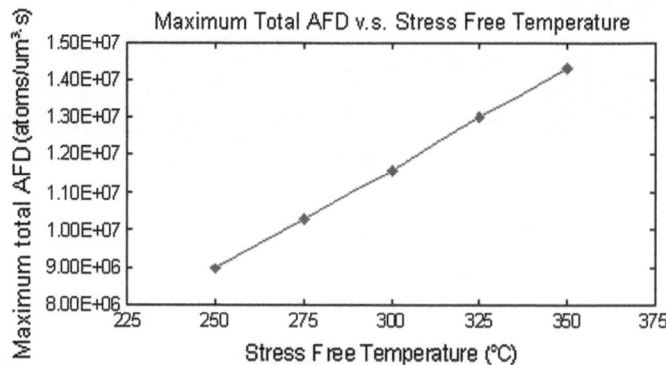

Fig. 4.24   Maximum total AFD versus stress-free temperature (Reprinted from [1], Copyright (2012), with permission from Elsevier)

increasing inter transistor distance is shown in Fig. 4.23. One can see from Fig. 4.23 that the impact becomes insignificant when the inter-transistor distance goes beyond 105 μm.

For Layout 4.21(b), as the original layout is already compact enough, any further reduction in the inter-transistor distance does not cause significant rise in the circuit temperature. The maximum total AFD of Layout 4.21(b) is similar to that of Layout 4.2.

4.4.4 Stress-Free Temperature of the Metallization

The high total AFD at the Via 23/Metal 3 and Via 56/Metal 6 interfaces is due to high thermo-mechanical stress gradient. Modification to the layout geometry is restricted by the design of the transistors. However, the EM lifetime is still able to be improved by decreasing the SFT of the metallization down to 250 °C [21]. When the difference between the interconnect temperature and the SFT drops from 265 to 165 °C, the intrinsic stress in the interconnects reduces proportionally. The stress built up and hence the stress gradient at the Via 23/Metal 3 and Via 56/Metal 6 interfaces are reduced. Figure 4.24 shows the decrease in the maximum total AFD with decreasing SFT, and it agrees with the simulation results of Roy et al. [22].

In summary, when the temperature gradient and thermo-mechanical stress gradient of the model decrease, a decrease in the total maximum AFD is observed. This reflects an enhancement in the EM lifetime for the modified layout.

4.5 Summary

In this chapter, the application of the 3D finite element circuit model to a realistic circuit consists of both intra- and inter-block interconnects up to the metal top was demonstrated, using a RF low noise amplifier (LNA) as an example. The analysis was first carried out on a full circuit model that contained all the circuit components and interconnections. As the simplified circuit geometry structures of the full model did affect the simulation accuracy, a sub-model simulation with the inclusion of the detailed circuit structures was performed at the problematic region as identified by the full model simulation.

The simulation results of the full model and sub-model showed great agreement, and the location of the maximum total AFD was found at the sharp turn of the source interconnect of Transistor M1 with the highest temperature gradient. With the inclusion of the detailed interconnect structures, the sub-model simulation was able to identify additional EM weak spots such as the interface between Via 23 and Metal 3 (i.e., the locations with geometry irregularity and thus high thermo-mechanical stress gradient). The locations with high thermo-mechanical stress gradient had comparably high AFD values as the locations with high current density and/or high temperature gradient. These locations were prone to EM failures and may be missed out if only current density was considered. This again proved the importance of including the thermo-mechanical stress effects into EM modeling and the inadequacy of the conventional current density based EM simulators.

The current density and temperature gradient at the metal top were found to be more than 10 times smaller than those at Metal 3 due to the larger thickness of the metal top, and thus the interconnects made of the thick metal layers were not a concern in this circuit model.

Modifications were performed based on the observations in the simulation, and up to 20.17–59.59 % improvement in the EM lifetime was achieved with the layout and the process modifications, respectively. Change in the orientation of the transistor might help to reduce the value of the maximum total AFD and enhance the EM lifetime. The results agreed with the experimental results in literature and thus validated the capability of the model for correlating the layout and process modifications with the EM reliability of an IC.

References

1. He F, Tan CM (2012) Electromigration reliability of interconnections in RF low noise amplifier circuit. Microelectron Reliab 52:446–454
2. He F, Tan CM (2012) Effect of IC layout on the reliability of CMOS amplifiers. Microelectron Reliab 52:1575–1580
3. Cheng Y-L, Lu Y-S, Chen S-A, Wang Y-L (2012) Effect of layout on electromigration characteristics in copper dual-damascene interconnects. J Vac Sci Tech B (in press)
4. Zutter D, Rogier H, Knockaert L, Sercu J (2007) Surface current modeling of the skin effect for on-chip interconnections. IEEE Trans Adv Packag 30(2):342–349
5. http://documentation.renesas.com/doc/YOUSYS/document/r09ds0009ej0100_microwave.pdf
6. http://www.hittite.com/content/documents/data_sheet/hmc318ms8g.pdf
7. Sungjun I, Banerjee K (2000) Full chip thermal analysis of planar (2-D) and vertically integrated (3-D) high performance ICs. In: Inter electron devices meeting. technical digest, pp 727–730
8. Federspiel X, Girault V, Ney D (2003) Effect of Joule heating on the determination of electromigration parameters. In: IEEE inter integrated reliab workshop final report, 2003, pp 139–142
9. Rzepka S, Banerjee K, Meusel E, Hu CM (1998) Characterization of self-heating in advanced VLSI interconnect lines based on thermal finite element simulation. IEEE Trans. Compon Packag Manuf Tech Part A 21(3):406–411
10. Shingubara S, Miyazaki S, Sakaue H, Takahagi T (2002) Evaluation of temperature rise due to Joule heating and preliminary investigation of its effect on electromigration reliability. In: AIP conference proceedings, vol 612. pp 94–104
11. Teng C, Cheng Y, Rosenbaum E, Kang S (1997) iTEM: a temperature-dependent electromigration reliability diagnosis tool. IEEE Trans. Comput Aided Des Integr Circuits Syst 16:882–893
12. Marín E (2010) Characteristic dimensions for heat transfer. Lat Am J Phys Educ 4(1):56–60
13. Schafft HA (1987) Thermal analysis of electromigration test structures. IEEE Trans Electron Devices ED 34:664–672
14. Chiang TY, Banerjee K, Saraswat KC (2000) Effect of via separation and low-κ dielectric material on the thermal characteristics of Cu interconnects. In: International electron devices meeting. Technical digest, 2000, pp 261–264
15. Lin MH, Chang KP, Su KC, Wang TH (2007) Effects of width scaling and layout variation on dual damascene copper interconnect electromigration. Microelectron Reliab 47(12):2100–2108
16. Vairagar AV, Mhaisalkar SG, Krishnamoorthy A (2004) Electromigration behavior of dual-damascene Cu interconnects-structure, width, and length dependences. Microelectron Reliab 44(5):747–754

17. Hau-Riege C, Klein R (2008) The effect of a width transition on the electromigration reliability of Cu interconnects. In: IEEE inter reliab physics symposium, 2008, pp 377–380
18. Darveaux R, Turlik I, Hwang L-T, Reisman A (1989) Thermal stress analysis of a multichip package design. IEEE Trans Compon Hybrids Manuf Tech 12(4):663–672
19. Osterman MD, Pecht M (1990) Placement for reliability and routability of convectively cooled PWBs. IEEE Trans Comput-Aided Design Integr Circuits Syst 9(7):734–744
20. Chao K-Y, Wong DF (2005) Thermal placement for high-performance multichip modules. In: Proceedings international conference on computer design: VLSI in computers and processors, 1995, pp 218–223
21. Lloyd JR, Lane MW, Liniger EG, Hu C-K, Shaw TM, Rosenberg R (2005) Electromigration and adhesion. IEEE Trans Device Mat Reliab 5(1):113–118
22. Roy A, Tan CM, Kumar R, Chen XT (2005) Effect of test condition and stress free temperature on the electromigration failure of Cu dual damascene submicron interconnect line-via test structures. Microelectron Reliab 45(9–11):1443–1448

Chapter 5
Concluding Remarks

5.1 Conclusions

With shrinking device size and interconnect dimension, electromigration (EM) in the interconnects had become the main failure mechanism that determined IC reliability. The conventional physics based 3D EM models were at the "localized" regions and thus could not represent the EM reliability of an entire circuit. As current density was no longer the sole factor that determined the EM reliability, and the temperature and thermo-mechanical stress distributions of the interconnects in an IC were greatly affected by the interconnect structures and the surrounding materials, the 2D EM circuit simulators based only on current density were no longer adequate. Thus, there is a need for 3D EM modeling at circuit layout level.

A new 3D EM modeling method at circuit layout level using the atomic flux divergence (AFD) approach was introduced in this work; 3D finite element circuit models were built from their 2D IC layouts, first for a simple two metal layer circuit with only the intra-block interconnects, and then for a complex circuit with the REAL circuit structures up to the metal top.

In Chap. 2, the method of constructing a 3D finite element circuit model was presented, using a simple inverter circuit as an example. Transient thermal-electric and structural-thermal analyses were performed on the circuit model using both Cadence (a circuit simulator) and ANSYS (a finite element software). The computed transient temperature response to the current and voltage changes and the transient thermo-mechanical stress response to the circuit temperature changes demonstrated the capability of the model for performing the transient temperature and stress analyses based on the activities executed by the circuit. This function was not available in the 3D models reported in literature.

The current density, temperature and temperature gradient, thermo-mechanical stress and stress gradient at any node of the 3D finite element circuit model were able to be extracted from the simulation results. The AFDs due to the three driving forces as well as the total AFD of the model were computed based on these values. The void nucleation location and nucleation time could be determined from the location and value of the maximum total AFD, respectively; therefore, the simulation results was useful in identifying the EM weak spots of the interconnects in an IC.

Models based on the inverter circuit using different barrier layer thicknesses and dielectric materials were simulated. An increase in the maximum total AFD was observed with a thinner barrier and with the use of the low-κ dielectric. The observations agreed with the experimental results in literature and hence demonstrated the structural and material modeling capability of the proposed model.

In Chap. 3, the EM performance of the circuit model (i.e., a circuit structure) was compared with the line-via test structure under different operation conditions. The output stage of a class-AB amplifier was used as an example. It was found that the dominant driving force for the total AFD was different at different temperature. At the EM test temperature of 300 °C, the dominant driving force for the total AFD was the AFD due to temperature gradient–induced driving force, and the locations of the maximum total AFD for both the circuit structure and the line-via test structure were at the place with the highest current density. However, at the circuit operation temperature of 90 °C, the AFD due to thermo-mechanical stress gradient–induced driving force provided more than 90 % contribution to the total AFD, and the current density was longer the main factor that affected EM. As such, the location of the maximum AFD was determined by the product of the thermo-mechanical stress gradient and the temperature gradient, and the EM weak spots of the circuit structure were different from that of the line-via test structure. The normal extrapolation using the line-via test structure was therefore unable to provide an accurate EM prediction for a circuit under circuit operation condition. This again implied the need for a complete 3D circuit modeling to ensure an accurate simulation result. Modifications to the interconnect structures of the model were performed, and the simulation results agreed with the observations in literature, indicating the accuracy of the proposed model.

In Chap. 4, a complex circuit structure up to the metal top was constructed with the inclusion of all other circuit components and both intra- and inter-block interconnects. A simple LNA circuit was used as an example. The simulation was conducted at both global (i.e., full model) and local (i.e., sub-model) levels. The full model modeled the EM reliability for the entire circuit and was able to identify the most severe failure site. However, due to the oversimplification in structure, the full model missed out other possible EM failure sites. This was a common problem in the 2D simulations for which the thermo-mechanical stress and stress gradient were overlooked due to structural simplification. The sub-model was used to solve this problem, and additional EM weak spots due to high thermo-mechanical stress gradient were identified.

Some common practices for enhancing the EM reliability of an IC were examined using the LNA model, and the results agreed well with the simulation and experimental results in literature. This demonstrated the capability of the proposed model for relating the layout and process modifications with the EM lifetime. The model could be conveniently used to predict the improvement in the EM lifetime based on the EM weak spots identified.

5.2 Recommendations for Future Work

This work is a beginning in the research field of the 3D finite element circuit modeling. Due to the complexity of the model and the constraint in computation memory, the microstructure and back stress effects are not included in the model. However, as EM progresses, the voids may move along the grain boundaries of the interconnects and nucleate at another location [1]. Furthermore, under the effect of back stress, the void may disappear from its original nucleation location. Therefore, it is important to include these effects. The dynamic simulation of the void nucleation with the consideration of the microstructure and back stress effects of the 3D circuit model is an interesting area for further study.

The examples used for the 3D model construction in this work are simple circuits containing only a few transistors. Practical circuits used are much more complicated and may consist of hundreds or thousands of transistors and other functional blocks. The way of laying the circuits in a chip and the interaction with other blocks may affect the circuit temperature and hence the circuit reliability. These factors are not considered in the current model. Work is being done by using the neural network method [2, 3] to speed up the computation, so that the proposed model will be able to be applied to large and complex circuits and the possibility of including it in the IC design flow.

References

1. Li W, Tan CM, Hou Y (2007) Dynamic simulation of electromigration in polycrystalline interconnect thin film using combined Monte Carlo algorithm and finite element modeling. J Appl Phys 101(10):104314
2. Zhang Q, Zhang L (2009) Neural network techniques for high-speed electronic component modeling. In: IEEE MTT-S inter microwave workshop series on signal integrity and high-speed interconnects, 2009, pp 69–72
3. Cao Y, Zhang Q (2010) Neural network techniques for fast parametric modeling of vias on multilayered circuit packages. In: IEEE electrical design of advanced packaging and systems Symposium, 2010